Halcyon River
Diaries

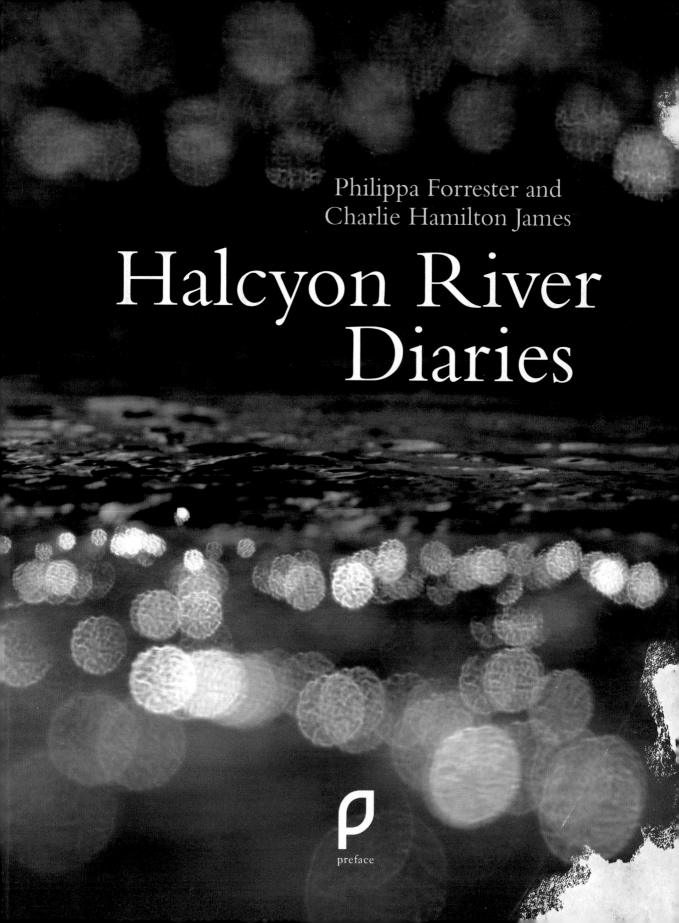

Philippa Forrester and
Charlie Hamilton James

Halcyon River
Diaries

preface

Mums – you inspired our love of the natural world and showed us the importance of a loving mother, this book is therefore for you.

HALCYON RIVER DIARIES
Philippa Forrester and Charlie Hamilton James

ρ
preface

Published by Preface 2010

Published to accompany the television series Halcyon River Diaries, first broadcast on the BBC in 2010.

BBC and the BBC logo are trademarks of the British Broadcasting Corporation and are used under licence. BBC logo © BBC 1996.

10 9 8 7 6 5 4 3 2 1

First published in Great Britain in 2010 by
Preface Publishing
20 Vauxhall Bridge Road
London SW1V 2SA

An imprint of The Random House Group Limited
www.rbooks.co.uk
www.prefacepublishing.co.uk

Addresses for companies within The Random House Group Limited can be found at www.randomhouse.co.uk

The Random House Group Limited Reg. No. 954009

A CIP catalogue record for this book is available from the British Library

ISBN 978 1 84809 225 9

The Random House Group Limited supports The Forest Stewardship Council (FSC), the leading international forest certification organisation. All our titles that are printed on Greenpeace-approved FSC-certified paper carry the FSC logo. Our paper procurement policy can be found at www.rbooks.co.uk/environment

FSC
Mixed Sources
Product group from well-managed forests and other controlled sources
www.fsc.org Cert no. SGS-COC-005091
© 1996 Forest Stewardship Council

Design: Craig Stevens and Nick Heal
Illustration: Jake Biggin

Printed and bound in Great Britain
by Butler, Tanner and Dennis Ltd

Contents

● Diary extract ● How to ● Practical ● Conservation ● Ten great places ● Recipe

Halcyon days

This is an extraordinary look at an ordinary river.

We were invited by the BBC to make a programme about a year in the life of a river we love. We live beside this river and Charlie has photographed the kingfishers on it since he was a boy. We are both wildlife filmmakers, Charlie does pictures and I do words and our children (Fred, Gus and Arthur) follow us around asking funny questions. When we agreed to make this series we wanted to include all of that, we wanted to inspire other people to enjoy the natural world with their families and reconnect.

When the river is your street and all the animals on it your neighbours you see them in a different way, get to know the everyday detail of their lives and discover new facts about them. But we wanted to use this year to push the boundaries of our own knowledge and to perhaps even make a positive difference to life here on the river.

We are not just observers but are part of the system; our bridges, our homes and us, all have an effect on the birds, animals and fish that use the water as their base and form this wonderful and fascinating ecosystem.

As wildlife filmmakers the obvious thing to do was to devote our expertise and equipment, our knowledge, proximity and every waking moment to filming the creatures, under, over and in and out of the water so that we could understand them better, inspire our children with a love of this natural world and ultimately learn how to look after this part of the countryside which is on our doorstep.

To be given this chance is a dream come true, I remember the thrill of anticipation of all the glorious filming days to come, what fun it would be with the children. It wasn't quite as idyllic as we imagined. An extract from my diary back then goes like this:

Today we have had a taste of what we are in for, namely that wrangling three kids and filming isn't half difficult. In blissful ignorance of this, Charlie and I sat on the bank watching Fred who was talking to the camera about making a nest box for dippers. My main worry had been whether the children would actually be any good with the camera, Fred had never really been that interested in performing, and like his Dad, would much rather take photos or film something himself, and you can't force a child to become a TV presenter if they just don't fancy it. However, he is really interested in dippers and because of that, he was communicating well and enjoying himself. Charlie and I exchanged grins, each knowing that the other was breathing a huge sigh of relief but we had exchanged a large amount of oxygen for carbon dioxide too soon. We soon realised that further downriver, in the background of shot, Gus had revealed his white bare bum to the world and with his trousers round his ankles was taking a wee; and Arthur had waded in after Gus, filled his boots with water and promptly decided that he was stuck. He began wailing, Fred was determined to carry on so simply talked more loudly and Jamie who was behind the camera was weeping in hysterics as he watched Gus. The camera wobbled.

After a few resets (starting from the beginning again), one of which ended in another welly disaster and then a couple of technical issues – the children were great, this time the microphone had interference – the moment began to wane, as indeed had the short winter afternoon.

Gus announced that he had ear ache, didn't want to do filming any more because it was boring and cold and stomped off indoors to watch cartoons. By now we had began to eat into Arthur's nap time so of course he became utterly miserable and with an expensive kit on hire I had no choice but to bribe him with sweets. He responded to the prospect of Marks and Spencer's Percy Pigs with a beatific smile, and happily pootled off to wade in the background as a contented child who lived on this idyllic river would, and so we started again. However, just as Fred was getting back into the swing of it, Arthur came slowly up behind him and then right around him, and just before Fred could finish, he peered around the camera lens, showed his empty hands covered in the remains of Percy Pigs and asked for some more sweeties giving the game away completely.

To top it off when we finally went to film the close-up of the fresh otter spraint which we had discovered earlier, we found Dave the dog licking his lips, having just eaten it.

This book is our chance to share our passion and inspire other families to get out there and enjoy our rivers and the wildlife. Even though it might not always seem practical, get the kids out, ignore their moaning and go for it. You will never regret it and neither will they.

Whether you want to just appreciate your local river, see more wildlife for yourself or get really close, this book will help you hone your field craft and photography skills, enabling you to get far more out of a visit to any river, and perhaps inspire you to make a positive difference to your local one.

Gus, Me, Fred, Charlie and Arthur.

Field craft

Charlie: **Most animals in Britain are inherently scared of us – they think we want to kill and eat them. Good field craft removes this problem by not letting them see us or dealing with them in a way that relaxes them and removes the threat, thereby allowing them to go about their business in a natural way.**

'To see without being seen' is a military motto that crosses over to describe the art of wildlife watching. The best wildlife footage and photographs are taken when the animals are not reacting to the photographer because he or she is using field craft to avoid disturbing them.

Several years ago I made a film on giant otters in Peru for the BBC. We were filming the otters on a small lake in the middle of the Peruvian Amazon. They always knew we were watching because we were in a boat right next to them. However, we used all our field craft to relax the otters so that they ignored us and went about their daily life with us just in the background. This field craft was nothing particularly special; it just involved keeping quiet, not standing up on the boat, always keeping a respectful distance between us and the otters, not invading their space but allowing them to invade ours, not making any sudden movements and most importantly watching them to constantly assess whether we were impacting negatively on them or not. The results of these few simple rules were that the otters learnt to trust us and allowed us incredible views of them bringing their cubs out for their first ever swim – something so rare and sensitive that we would never have got it if we had not followed the strict rules that we put in place.

The same set of rules is true whether you're filming giant otters in the Amazon or blue tits in your back garden – respect the animal by being quiet and careful

Respect the animal by being quiet and careful and you may be rewarded with views you would otherwise not get.

and you may be rewarded with views you would otherwise not get. Field craft is a vast subject and every animal needs to be dealt with in a particular way if your aim is to not disturb it. It is something that needs to be learnt by experience and time in the field. I can however suggest some key rules to getting close to wildlife.

Rule 1 – Know your subject

This is vitally important and you should try to find out everything you can about

your subject before approaching it. This knowledge will help you immeasurably when working with the animal. Knowing what its key senses are will have an effect on how it's best approached. A river otter, for instance, has appalling eyesight but it does have highly acute hearing and sense of smell. I've had otters asleep by my feet. The wind was in my favour and I didn't move an inch so they had no idea I was there. Such an encounter wouldn't work with a wild cat. It would have spotted me long before it got to my feet. So each animal requires different forms of approach and, as the approacher, you need to know the strengths and weaknesses of that animal's senses. Also the more you know about an animal the more you can interpret its behaviour.

Rule 2 – Stay quiet

Most animals react to sound and most are surprisingly good at telling which sounds are coming from humans and which are not. Most mammals in the

UK have hearing far more acute than our own and they can react to us and vanish long before we have even become aware of them. So stay quiet. If you're with other people, whisper. If your phone is on put it on silent mode. If you're wearing outdoor clothing make sure it doesn't rustle. Keeping quiet not only allows you a greater chance of seeing your target species but also of encountering many others.

Rule 3 – Stay low

Animals and birds have various 'search patterns' that they use to distinguish us and other animals. The most obvious search pattern is spotting the outline of a human figure – a silhouette of a human standing on a hill, for instance, would set alarm bells ringing in the hearts of most animals. So this must be avoided. When waiting for an animal such as a deer, stay low. Sit down and wait. The moment you get low the chance that anything will see you is reduced dramatically. It's the same

I continue to talk to let them know I'm a human. They generally then leg it, however they leg it with a lot less fear than they would have had if I'd allowed them to tread on my toes.

Rule 5 – Watch your subject

Watching the way an animal responds to you once it has seen you is not difficult. Most animals will run away, however some will not and you can make these animals feel more relaxed with a few simple techniques. Firstly, once they know you're watching them, pretend that you're not. If you are lying low and fixed on them they'll think you're stalking them. The best thing to do is to make them aware that you know you've been spotted. It's the lion walking through the zebra herd phenomenon. If the lion is relaxed and walking boldly, the zebras know that it is not hunting, it has no element of surprise – lions are ambush hunters. The zebras respond by not running from the lion. The same is true for many other animals. If you let them know you're there then they'll put up with you as long as you don't get too close. I remember years ago stalking an otter on a beach in Shetland for ages. It had caught a fish and I was snapping away with my camera about thirty feet from it, lying low on the shore. The local sheep farmer appeared, walking along the beach, and stopped to ask what I was doing. The otter looked at him and snorted in disgust. It didn't swim away, though, and so I stood up and the farmer and I chatted for five minutes or so. The otter continued to eat its fish and after the farmer had gone I continued to photograph it. Instances like that are rare but they do happen. My feeling is that the otter would have been more scared if it thought I was stalking it than if I was just standing there having a chat.

with the military – if you're going to wander around in no-man's land you're going to get spotted.

Rule 4 – Don't invade an animal's space

Getting too close to an animal is not only going to disturb it but will ruin your chances of getting a view of decent natural behaviour. The best thing to do is stay back. Ideally don't approach it at all but stay and wait in one spot, and allow the animal to approach you. If you're out in the sticks the subject is probably not expecting to see you and so may well be off-guard and relaxed. Be respectful though. I regularly have close encounters with animals such as foxes, deer, badgers and otters if I'm waiting quietly. My general rule is that when they get too close I talk to them, very gently and quietly, which alerts them that they are approaching a possible threat. When they hear me they usually stop and size me up.

Eye contact is also very important. Animals will watch your eyes to see if you're watching them. Staring at an animal will spook it. It's better to pretend you're looking the other way. I once encountered a cormorant sitting on a rock on the shore, also in Shetland. I wanted to get a photograph so I began approaching it. It watched me very intently. I watched it back but surreptitiously – head facing away. I moved towards it very slowly. When I was about fifty feet from it, it became nervous and looked like it was going to fly. So I stopped moving and sat motionless, looking the other way. When it settled I started again. Half an hour later I was sitting three feet from the cormorant and snapping away with my camera. I had basically stuck to the same routine – get closer then when it looks nervous stop and look the other way. It was a real breakthrough for me and I learned a lot from the experience. I later went on to use this technique to catch oiled shags and cormorants on the beaches in Shetland and Galicia after two major oil spills. I also use it every day at home. There are almost always moorhens in the mill pool outside our house. They are shy and timid and usually scoot off if I cross the bridge. They are much more relaxed, though, and will often continue about their business if I don't look directly at them.

Field craft is a great skill to acquire. It is rare to ever truly master it as it is such a vast subject. But learning to understand animals and the way they think will allow you much better views of them. One of the best ways to start is simply to find a nice quiet spot, get yourself comfortable and just sit quietly – you'll have to wait for a while but you'll be amazed at what pops out of the undergrowth to take a relaxed look at you.

If you want to really watch animals at close range the best thing to do is simply remove yourself from their view. The following chapter on hides and camouflage should give you a few pointers.

Hides and camouflage

Charlie: I hate sitting in hides. Unfortunately I have no choice if I want to get close to wildlife. Hides come in all shapes and sizes but all essentially do the same thing – they prevent the animal that you are watching from seeing you.

Hides can be made from any material and be any size. The ones at the Wildfowl Wetlands Trust at Slimbrige are large and spacious, allowing you to walk around and watch the birds through small slit windows. The birds there are so used to the people and hides that they completely ignore them. The kingfisher hide is by far the best, not only is it beautifully designed and made but it looks out over a small pool and a kingfisher nest bank. In the spring and summer you can stand there and watch the full life cycle of the kingfisher in glorious technicolor.

My own hides, however, are far from luxurious. I generally use the same camouflage dome hide that I've had for years. It is about four feet square and about five feet high. It has a large window at the front and one small window on each side. The back Velcros and ties up to seal me in and when it rains water drips through the roof onto my head and camera. But it does work, so well that the birds I want to film from it often sit on the top, denying me the all important view of them. These dome hides are very useful. They pack up small and light and take seconds to put up. Back in the old days we used square box hides made from thick green material with elaborate tent pole systems. They were rubbish. They took ages to assemble because the poles would keep collapsing. By the time you'd actually got it to stand up, you'd be so frustrated you'd want to smash it up with a cricket bat.

When I was a kid, finding a decent hide was a nightmare. I couldn't afford to buy one so I had to make do with whatever I could. My mum attempted to make me one. We used to live next door to Cameron Balloons who made hot air balloons and we would raid the bins at night for all the rip-stop nylon that had been chucked out. Mum and I gathered a fair pile of it and together with a load of bamboo sticks constructed a reasonable hide. It didn't last long, a few trips to the river and it had all but disintegrated. I once made a hide out of a bed sheet. I was on holiday in the Algarve with my

Hides are excellent for getting close to animals but you still need to know what you're doing with them.

parents and found a bee-eater colony near our hotel. So I spent the holiday photographing the bee-eaters from beneath the sheet, using a coat hanger to rest my camera on. It worked and I managed to get some great shots of a Montpellier snake raiding the nest colony. The best hide, however, was an old Landover tarpaulin. It was thick, green and waterproof with a large plastic window. It was rigid enough to almost stand up without poles and after cutting a couple of holes with my penknife it was perfect. This lasted me for years of kingfisher-watching – until it finally rotted.

There are lots of different types of hide on the market at the moment. Some have built-in chairs, others collapse and erect in one go. Most are too bulky and complicated and have windows in the wrong place. The problem is that many are designed for shooting animals with guns rather than cameras. I bought my first dome hide about ten years ago and I have since failed to find a better system. It does everything I need. It could do with being refined but essentially it works.

Hides are excellent for getting close to animals but you still need to know what you're doing with them. If you move around too much, make a noise, or don't conceal yourself within the hide properly you are going to blow your cover. The key is, hides allow you a little more freedom to move but not too much. You need to stay still, keep quiet and cover your lens with scrim so that you're not spotted. Some birds, especially birds of prey, are very alert, watching for the slightest movement. The kingfishers I work with will even watch the lens elements move as I zoom. If I move the lens when they are watching they'll spook, so I have to make sure they are relaxed and preferably looking the other way before I adjust my camera. When looking out of a hide, never stick your eye up to the window. Instead stay back a bit in the shadow. You rely on being darker than your subject; if light is shining in and lighting you, you may well be spotted.

Positioning a hide is also important. Generally I try not to get too close to my subjects. I want them to behave naturally and although they might be unfazed by the hide, they may well react to the lens and its movement. So I have to balance between being close enough to get the close-up shots and being far away enough to ensure the animal is relaxed. Some wildlife photographers insist that hides should be moved in very slowly so as not to disturb the animal. This does make sense with some species. In my experience there are few birds that are concerned with a hide if it is more than thirty feet away from them. For me, the key is not so much the hide but how the person within is operating. If you stay back, stay quiet, don't move much and above all respect your subject then you shouldn't have any problems.

The biggest problem I have with hides is the waiting. Extreme boredom can be seen as an occupational hazard. Sometimes I get so bored it hurts! The arrival of the iPhone helped massively (sad as it is). Now I can sit quite happily watching movies and drinking coffee while I wait. Obviously I may miss things as a result but luckily my camera has a 'pre-record' – which records 8 seconds before I even hit the button!

Camouflage

Dressing up like a Royal Marine is another great way to get close to wildlife. I used to rubbish the idea and as far as I was concerned people who dressed up in camouflage were just taking themselves too seriously. Indeed on one shoot I was doing for the BBC I deliberately wore a fluorescent orange jacket to film otters. It worked so well that the otter ended up eating a crab at my feet for ten minutes. What I didn't know at the time was that bright orange is a very common colour used in camouflage clothes. Simon King re-educated me in this area. He explained the benefits of camouflage and I gave it a go. The results were fantastic. I got myself kitted out in camouflage everything and set off to film foxes and deer. I couldn't believe it, they just didn't see me!

Camouflage comes in different patterns and colours. The more modern designs such as Advantage Leaf Pattern from the US mimic natural surroundings far better than the classic old British army camouflage. Some of these clothes are amazing feats of design, with specific species of leaves and grasses printed into the material. The more expensive jackets are also fully-waterproof and are made from special fabric that doesn't make a

> The kingfishers I work with will even watch the lens elements move as I zoom.

noise when it rubs against itself. Combine these jackets with similar patterned trousers, shoes, gloves, hat, face mask and camera bag and you literally become invisible (except of course when you're standing in the kitchen showing off your new garb to your uninterested wife).

If you do make the effort and really go for it with the camouflage, you can get very close to animals. Many birds and animals recognise us by our various forms – a standing silhouette of a person is instantly recognisable by just about any animal. The more we break up that form the harder animals find it to recognise us. I regularly have kingfishers sitting right next to me if I am crouching down in the river doing something. The bird soon works out what I am and flies off, but it is fooled by me briefly because it doesn't realise what I am. Camouflage is simply breaking up the human form, blending it in with the surroundings, making it difficult or impossible for animals to see us.

Some camouflage also relies on the senses of certain animals. Deer, for instance, have poor colour vision and cannot see orange. Bright orange is therefore used for hunting deer in the US. Not only does it appear to be a neutral dull colour to deer but it allows hunters to see each other, which in a country where people have a habit of shooting each other is arguably a good thing. I found the use of orange very useful when I was in India filming smooth-coated otters for BBC's *Planet Earth*. Unlike our otters, the ones here had very good eyesight and so I was wearing full camouflage in order to get close to them. I had a spotter on the other side of the river but he and I were really struggling to see each other as he was camouflaged too. We both had camouflage hats, though, and the insides of them were bright orange;

so we turned our hats inside out and stood out like sore thumbs. The otters, however, couldn't see the orange so we were able to spot each other and keep track of the animals with ease.

The big problem with both camouflage and hides is of course scent. Most mammals have good sense of smell and no amount of canvas or clothing can prevent them smelling us if we're upwind of them. This is being addressed by various companies who are now making clothing that uses carbon to absorb human scent. Other companies are manufacturing badges that you stick to your clothes which mask the smell of humans by giving off the overpowering smell of rotten logs. (Note from wife: this is *not* pleasant.) There are all kinds of products to mask our smell these days – whether or not any of them work remains a mystery to me – especially when I've been eating lots of baked beans.

Halcyon River Diaries

Early Spring

Philippa: The spring flowers are putting on quite a display this year. It seems as if they are, in all their different coloured forms, nodding their approval along with us humans that we have finally had a decent winter with a proper cold snap.

The stuff that Christmas cards are made of: apple-cheeked toddlers in mittens and snowmen with scarves on. As if to remind us of how good it was, snowdrops are all over the ground in luminescent white pools looking leggy and lovely, crocuses bring colour and a reminder that there are more exotic plants waiting underground for the months to come.

My bed of heligan hellebores that I have nurtured through the last few years seems to have heard the call to action and are thoroughly committed to the show, a troupe of belle bells the best they have ever been.

Then when my eye is distracted upwards by the blue sky I realise that the catkins are also glorious this year. I can't ever remember noticing them so clearly before, but their yellow festoons break up the hedges and are already giving some colour in the treetops. But the daffodils are stealing the show and are really the most dazzling dancing girls on the stage. This year their choreography has been impeccable. The last few years, to be frank, have been shoddy – a few confused flowers dazily putting out their petals in December whilst the Christmas decorations were up, a few in January and then a final half-hearted flurry in February. It was clear that none of them really knew what they were doing and as a troupe they were all over the place, as I say, shoddy.

This year there has obviously been a concerted effort to co-ordinate the

Apple blossom brings a splash of colour to the garden.

The daffodils are stealing the show and are really the most dazzling dancing girls on the stage.

upward explosion from the bulb. They have all burst into flower in unison and the display is dazzling. All flowers – be they white with orange centres, simple yellow or fussy double frou-frou orange – are standing tall in natural-looking clumps that belie the amount of effort such an enthralling dance has taken. I applaud them every time I walk up and down the path beside the stream, they deserve it.

The stage of my riverside garden has been bleak for long enough and these fantastic creatures which are dancing onto it with such determination put a smile on my face. I am certaint that somewhere beyond us something deep in the life blood of this place is singing in the knowledge that this year is going to be great.

Halcyon R

How to spot a kingfisher

Charlie: Most people who have seen a kingfisher can remember when and where they saw it. Kingfishers stand out from other British birds because of their exotic, almost tropical colour.

Generally kingfishers in Britain are shy and elusive and most sightings of them are just flashes of blue as they zip past at high speed. But if you know where and how to spot them you stand a chance of getting a much better look.

Most healthy lowland rivers in England and Wales have kingfishers living on them. Ireland does pretty well and so do parts of southern Scotland (kingfishers are not present in much of Scotland though).Whether the river is in the centre of the countryside or in the middle of the city is not important – I once saw a kingfisher flying through Bath Spa station while waiting for a train. Kingfishers generally prefer slow flowing rivers with overhanging vegetation and a good supply of fish, so you can improve your chances even more by finding one of these.

Once you've found your location, you need to sit and wait. If it is in the middle of a city, the kingfisher will have lost its normal shyness of humans and will often come quite close. So find a bridge or riverside bench and be patient. If you're out in the sticks your best bet is either to watch from a bridge or to try sitting in some cover – if you can break up your shape the birds may not recognise you as a human or threat, as long as you keep very still. I have had hundreds of very close encounters with kingfishers without any cover at all. Not moving is key. Kingfishers have incredibly good eyesight and they will lock on to and study the slightest movement, so if you twitch

Good camouflage can work very effectively and allow for some very close encounters.

you could get caught. Dull green or brown clothing also helps and completely blending in with good camouflage can work very effectively and allow for some very close encounters.

Field signs are also very helpful. Kingfishers are creatures of habit so reading the signs they leave can improve your chances of a really good sighting. A pair of birds patrol a territory of around a mile. Within that stretch they will have several favourite fishing spots. Often they will have a favourite perch at these spots and they will do almost all their fishing from it. These perches are not that hard to find as the kingfishers will defecate while sitting at them and leave a spray of bright white around them. If the perch is above ground rather than directly over the water, check for pellets. Pellets are made of fish bones that the kingfishers can't digest and so regurgitate. They are almost white and about the size of a small broad bean. If you find what looks like a favourite perch, sit and watch it for an hour or so – if it is, the kingfishers will probably appear and fish from it for a while.

↓
Pellets beneath a perch are a sure sign that kingfishers have been using it.

Halcyon River

By far the best way to spot a kingfisher is to know what they sound like. Kingfishers may be small but they are loud. Their main call is a series of high-pitched whistles blasted out as they speed up and down the river. They use the call to announce to other kingfishers that they are coming and this is their territory. If you know what it sounds like your chances are greatly improved. I suggest going on the internet and finding a recording of it to familiarise yourself with before going out looking.

Once you have found your kingfisher you may want to take everything one step further and get a hide in place to watch it from. Kingfishers are not bothered by hides, although if they see movement from one they will get spooked. I have read countless tales by wildlife photographers about how they had to carefully move their hides closer to the birds each day so as not to disturb them – this is certainly true in some instances and with some birds, but I often think these guys are over-egging it a bit. A hide in a river thirty or so feet from a kingfisher's perch will not bother it in the slightest, unless it sees movement within it.

Kingfishers are actually fairly easy to work with as they are such creatures of habit. So be patient and persistent. Get comfortable and wait. You'll be amazed what other animals you may see while you're waiting for them.

↓
Male kingfishers have black beaks – females an orange lower beak.

Halcyon R

Charlie: Dippers are one of my favourite waterside birds. They look dull at first glance – just white chest and brown back – but closer inspection reveals a stunningly intricate pattern of brown feathers, white eyelids and a chest so dazzling that it stands out from everything else along the river – making the dippers easy to spot.

Dippers are terrific characters. These feisty little birds lead complex and interesting social lives – something I wasn't really aware of until I started spending time with them.

It's dawn. I'm in my hide with my brother Stephen. The hide is in the river and we're sitting with our booted feet in the water, so we're both shivering with the February cold. We're out early to try and record the sound of the dipper singing. There are two dippers working the patch where I've put the hide and they are courting each other – which means lots of singing.

Stephen is our sound recordist for the day – I have tempted him out of bed so early by telling him that few people have ever recorded the dippers' song and no one recently on modern kit. Stephen is wishing he was still tucked in as he uses his radio mic to try and record the song. A radio mic or 'tie mic' is usually clipped on to a person's tie or shirt collar to record them when they are talking. It is very small and attached by a wire to a transmitter which sends the signal to the sound man wirelessly, which he then records. Radio mics only really work well if they are very close to the subject. So Stephen has placed it on a small gravel beach opposite the hide in the hope that the dippers will sing next to it. Of course the dippers could land anywhere and so our chances of getting them within range are slim, but we're going on the mantra 'nothing ventured, nothing gained'.

Dippers do have favourite hunting spots — I know that this beach is one of them and it's not long until one turns up and lands right by the mic. It hops around and starts poking about in the river. I film it – it is not singing though. We wait with nervous anticipation, Stephen clasps his headphones tight to his ears. Of course, it doesn't sing!

Rush hour on our country lane is not quite the same as it is in many other places, however the river runs alongside the road and the morning rush does mean a car or so a minute travelling past. This might not sound like much but when you are recording sound it drives you crazy. We have also ended up directly under the flight path of the local airport, so we're almost always waiting for planes to pass.

After a short burst of foraging the bird lands back next to the mic and dips for a moment. I hear the distinctive repetitive clicking call of another dipper flying down-river towards us. The dipper by the mic immediately starts singing, Stephen gets very excited and starts twiddling knobs on his sound mixer. He goes quietly nuts as a car rumbles past, drowning out the song completely.

The dippers meet each other on the beach and the male fans his wings out and hops up to the female, wings shivering, they seem to dance together. He has some food in his beak and he

tripod on the stony beach and wired back to Stephen's sound mixer, which in turn is wired into the back of my camera. This means that the sound Stephen is recording will be laid down on to whatever I am filming – hopefully a dipper singing!

The rig is much larger and more intrusive than the radio mic but if it works it will give us perfect sound quality.

The dippers are busy too. They seem to be building a nest in the underside of the stone bridge downriver from us so they are zipping backwards and forwards past the hide. After an hour or so one of the dippers lands back on the beach and pokes about in the stones. It moves closer and closer to the stereo mic and I start filming it. As it reaches the base of the tripod another dipper appears out of nowhere and lands right on the mic! The two birds then sing to each other, one on top of the mic and one underneath. Stephen can't believe his luck and I try to get as many different shots of it as I can. The weird whirring, popping and fluting goes on for about forty seconds until a car drives past and ruins the sound, the birds then split and fly off to the nest. We got it though!

Stephen gives me the headphones and I play back the sound through my camera – it is stunning. We go back to the house happy that we have pulled it off. I have been bitten by the dipper bug now and I want to spend more time watching them and getting to know more about them.

passes it to her, she accepts it and he hops back and continues to sing.

A car passes and Stephen does not look happy. He's lost most of the song to the the vehicle noise, just getting a tiny bit at the beginning and end. We stay in the hide all morning until the noise level by the river becomes unworkable.

The next morning we're out even earlier, trying to beat the rush. Stephen has changed tack and tries using a stereo mic. This is quite an elaborate set-up – two long mics (about a foot long each) suspended by small orange rubber ...ds, one above the other in a cage ...ng. The mic is then attached to a small

Dippers singing

If you want to hear dippers singing try looking on the internet. Here's a link for Radio 4 **http://www.bbc.co.uk/radio4/ science/birdsong.shtml**

Philippa: Today Arthur and I take a walk in the park – the river there is much bigger than ours and there is a lovely footpath all the way along it, although under the bridges is the typical graffiti.

Arthur zooms along on his like-a-bike, little legs striding out, and I get a power walk, big legs striding out. There is the usual collection of drab mallards hanging out like youths at a bus stop waiting for hand-outs. Somehow in gangs in the park, they aren't as glossy.

Further down the path where the river is broader we spot a heron by the water's edge on the opposite bank and stop for a breather. I point him out to Arthur. He is hunting, locked on to something under the water. Suddenly the spear of a beak is in and out, grasping a writhing eel.

is really writhing, it's not nice to watch. It has wrapped itself right around the heron's beak. I'm not good with snakes or eel-like things and so am not particularly relishing the sight but the heron is completely unperturbed and simply ignores it.

'It's a snake.'

'No, it's an eel, they live underwater, I wonder if we've got any in our river.'

I know we used to have them, our old next door neighbour had told us about them and I saw one once a few years ago, but nothing since. You'd probably

The eel thrashes with more vigour, sensing that escape into the water is just a few yards away …

'Ooooh! Look what he's got,' I say – any passer-by overhearing would have recognised the exaggerated over-excited manner with which you find yourself speaking to a toddler, but actually I am surprised. An eel is the last thing I thought the heron would impale.

'It's not often you see eels now.'

'No!' Arthur says, using exactly the right intonation although he doesn't really have a clue what I am talking about.

'Look, the heron's got an eel,' I crouch down and put my arm around him.

'Mmmmm!'

'Can you see it?'

'Yes, it's wriggling – eugh!'

And I know what he means – the eel

expect to see the remains of them every now and again since they are an otter's favourite food. This is quite a large eel so must have been around in the river for a while.

'It's dropped,' Arthur laughs.

The heron has indeed dropped the eel on to the muddy beach where it continues to thrash around. I'm surprised, herons have always struck me as efficient hunters. Maybe this eel had been uncomfortable, tightly-wrapped around his beak like that.

The eel thrashes with more vigour, sensing that escape into the water is just a few yards away. It manages to co-ordinate itself and heads in the right direction. We look on, so does the heron. We will the eel on, the heron just watches it escape. And then just as the eel tastes the water, the heron takes one long-legged step and lunges, then is still again, just standing there on the muddy bank like an old gentleman in a club trying desperately to be dignified and ignore the fact that his moustache is wriggling.

Arthur laughs again, warming to the heron's antics, 'He got it!'

'Mmmmmm!' He glances up to read my expression, sensing my disapproval. Even though its squirming makes me shudder I am starting to feel sorry for the eel.

'It's dropped again,' Arthur squeals, delighted at the repetition.

Even the eel is getting used to the idea and is much quicker off the mark this time, but no sooner has its nose touched the water than, pow, the cruel beak is there. It happens again and then again, but this time the heron nearly loses out. The eel manages to disappear under the water and must be more difficult to grasp – the heron picks it up but drops it again into the water, nearly tripping over his long legs trying to keep up with it.

'Gone!'

'Yep! Serves him right,' I say.

'He's got it again.'

'Oh yes, so he has.'

The heron, at this near-miss finally decides enough is enough and promptly starts to swallow the still-wriggling eel.

'He's eating it.' Arthur is now enthralled.

It is quite the most disgusting thing I have seen in a long while.

'It's still wiggling!'

'Yes, I can see that.'

The eel is now almost half gone – only the latter half is wrapped around the heron's beak but it's still trying to hold on and we can see the first half still

wriggling inside the heron's long neck, causing it to bulge out. Despite having atrocious table manners, the heron retains its composure, as if this was the done thing at the club.

With every large gulp the latter half loses a little more grip until the final one signals our last glimpse of the eel's tail. If the heron had a napkin he would now have delicately dabbed the corners of his mouth, not a splash on his white front, he just stands there.

'He's eaten it!' says Arthur. Show over, he prepares to mount his bike.

'Eaten but not defeated,' But he is no longer listening to me, as I slowly stand I still can't take my eyes off the heron – repulsive as it is, part of me is fascinated. The eel may be out of sight but it is still wriggling and I can clearly see it inside the heron's neck.

But my son has gone on and I need to start power-walking to keep up. Intrigued as I am, I can't stand and wait to the final wriggle. I can't help but wonder, though, how long it will take for the heron's digestive juices to take effect.

I begin to think, if there are big eels feeding the herons on this local river, there might still be eels on ours, but how do we find out?

Charlie: It's dawn and I'm in my hide at the far end of the kingfishers' territory. The sky is dull and grey and my camera is struggling for light. I've positioned the hide on a small shingle beach about twenty metres upriver from a nest bank that the birds are taking a very keen interest in.

The bank is about eight feet high, steep at the bottom and inclined towards the top, with lots of overhanging vegetation – a perfect place for kingfishers to nest.

Both birds are sitting together on a stick that hangs from the roots of an ash tree outside their nest. The male has a lump of mud on the end of his beak and the female also has a dirty beak. This means

only one thing, they've been digging a nest. Kingfishers dig their nests into river banks by stabbing and excavating with their sharp beaks. The nests can be up to a metre long with a chamber at the end, and can take up to three weeks to dig. These birds seem to be doing quite well. I watch the female fly in and vanish into the hole, staying there for nearly five minutes. Occasionally she ejects the mud she has dug and it trickles down the bank and splatters in the water.

This is one of the most active periods in the kingfishers' year. Not only do the birds have to jointly decide on where they are going to nest but they also have to dig it, bond as a pair and then eventually mate. For this reason they start their breeding season quite early, often in the middle of February, although their first brood of chicks may well not hatch until the beginning of May. This pair seem to have finally settled on a nest bank and are now digging hard, but it has taken them almost a month to decide on the best spot – they nearly nested in the bank outside our house, but after a couple of days of digging, changed their minds. This nest bank may be better for them, though, it is higher and more inaccessible to predators such as mink and rats.

As well as picking up shots of the birds flying in and out of the nest hole, I am here to film a fish pass. This is a key courtship moment and occurs at the time of the breeding season when the pair are cementing their bond. It is the male who initiates fish passing. He catches a fish and then flies over to the female, sits next

to her on the perch and offers it, head first, as a gift. Early in the season the female will very often ignore his advances and catch her own fish. And the male will offer her a fish which he wouldn't eat himself, such as a stone loach (a small slimy fish generally avoided by kingfishers). It seems that the female likes to test the male before she succumbs to his fish passing and finally accepts one. After a few successful passes, the male becomes a little blasé and lazy. I have watched on many occasions female kingfishers begging the males to catch them a fish and the male birds completely ignoring them, even catching fish and eating them in front of their female mates.

Filming a fish pass is not easy and that is why I'm in my hide so early this morning. I'm not only waiting for a fairly uncommon piece of behaviour but I'm hoping that it will happen right in front of me so that I can get a clear shot of it.

I've been watching the pair for nearly an hour. They are taking turns going in and out of the nest, each for several minutes at a time. The male bird eventually flies off downriver leaving the female in the nest. I assume he's gone off fishing. It always amazes me watching kingfishers digging nests – they are serious grafters. They do have to stop every hour or so for a fish snack, though. After a few minutes the female pops back out of the nest and lands on the perch. She calls to the male with a couple of high-pitched peeps but he doesn't respond. She scans the water for a fish but none pass below her. She cleans her beak by scraping it along the branch as if sharpening a knife, then sits for a while.

A series of high-pitched whistles heading upriver announce the male's return. The female tucks her feathers into her body and leans her head forward, calling him as he arrives. As he lands the two both adopt an aggressive posture towards each other – feathers drawn tight to their bodies, beaks open, heads up, tails splayed. But a few seconds of whistling to each other and they relax. The male has a juicy fat minnow in his beak. I turn my camera onto him, zoom in a bit and get my focus; my heart is pounding. I get very nervous when I'm anticipating getting a shot and this could be it. I hit Record. The male turns around and shuffles down the perch towards the female until he is sat right next to her. He offers her the fish, whistling constantly. She turns away and shuffles off a few inches. The male bird follows her and leans in with his fish, again offering it by touching it against the end of her beak. She eyes the fish for a moment and, encouraged by the male's constant whistling, opens her beak and tries to take it. Of course, the male being the male, he isn't actually sure that he wants to release it! So the two birds enter into a tug-of-war, beaks both locked onto the fish, for nearly thirty seconds until the female finally gets hold of the poor fish and sits up with it in her beak. The male bird then sits bolt upright, splays his tail feathers and then flies up in a sort of weird arc above the female and then off downriver, whistling excitedly.

I get the shot. I keep the lens focused on the female for a few seconds while she downs her minnow. Then I turn Record off and sit back in my chair a happy man. Now I just have to film them mating – that's a lot harder!

Charlie: Otters are surprisingly easy to track. They live in linear habitats – along rivers or stretches of coastline. Other animals, suchas foxes and badgers, roam over much larger areas so finding their signs can be harder.

Favourite spraint rocks turn green from years of 'fertilisation'.

Otters like to spraint, or poo, as the boys call it, regularly on particular spots. Otter 'spraint' is the number-one piece of evidence to find when looking for otters. Indeed, many of Britain's otter surveys are carried out by people who never actually see otters, they just find their spraints.

So what does a spraint look like? Well, to be honest (and not surprisingly) it looks like a small pile of poo. It varies in colour from brown to green, black or even orange if the otter has been eating signal crayfish; in consistency from sludgy to hard; and in smell too. Smelling spraint is something us otter people do a lot of. It is the definitive way of differentiating it from that of other mammals because, rather surprisingly, otter spraint actually smells quite nice! I regularly walk along rivers with people and pick up otter spraint which I insist they sniff. Nobody ever wants to but everybody who does has to admit that it doesn't really smell that bad. The general view is that it smells like jasmine tea. But that doesn't mean you should drop a piece of spraint into a mug of boiling water and allow it to stew for four minutes. I have never smelled jasmine tea – to me otter spraint smells like someone's put an Earl Grey teabag in an ashtray. Anyway, the bottom line is that, unlike fox, mink and other mammals whose poo really smells like poo, otters poo perfume.

To me otter spraint smells like someone's put an Earl Grey teabag in an ashtray.

Otters tend to lay their spraint on prominent marker points. A large rock in the middle of the river or a bridge footing would be perfect. The idea is that they are leaving a marker to other otters, saying, 'I live here and this is my profile' – young, old, male or female. As otters are fairly territorial creatures, this kind of information is very important to them. Otter spraint not only tells us that otters are present on a river but it lets us know *how* present. Areas with tiny amounts of very dry spraint can suggest that otters have passed through or are present but in low density. Fresh spraint still has bubbles in it and looks wet; if you're really hot on its heels you can sometimes see the water patches left by the otter as it climbed out of the river. The otter hunters of old were experts at reading spraint. They could tell the size and sex

↓
A classic mid-river
sprainting rock.

of the otter which left it and which direction it was travelling. I have to confess I have never become this accomplished, although I could probably hazard a few guesses.

Footprints are the second most important clue when tracking otters. They are not difficult to find because most rivers have muddy or sandy edges. Otters spend a lot of time getting in and out of the river when they are foraging and so tracks will not be too hard to find if they are in the area. To the trained eye the tracks are very distinctive. There are usually four, sometimes five, toes which look like upside down tear drops, splayed out from a central pad. A good set of tracks in soft, shallow mud or sand will give an accurate impression of the pad and the distance between the footsteps, so one can get a pretty good idea of the size of the animal. Tracks are also a great way of finding out whether cubs are present on the river, as obviously the sizes of footprints will be very different. On rivers otters don't zigzag about, they tend to travel in one direction so a set of tracks will indicate which way they are going. You can also gain some idea of the frequency an area is used by the 'padding' found there.

Otters will frequently hunt ponds and lakes adjacent to the rivers where they live and are particularly fond of those with fish

Tracks are a great way of finding out whether cubs are present on the river, as the sizes of footprints will be very different.

↓

It's easy to tell what an otter has been eating by examining the spraint for fish scales, bones and crayfish remains.

and frogs in them. If they use these lakes regularly they will almost certainly leave a trail from the river to the lake. These are usually fairly direct and thin trails. At either end will be an area flattened down as the otter leaves the land and slides back into the water. I particularly like tracking otters in snow. It never ceases to amaze me just how far and wide they will roam – snow allows us the rare chance to see this. I have found tracks from otters going for miles across hills in Shetland and through fields and farmland in England.

The third good clue to the presence of otters is dead fish. Otters eat different parts of various types of fish. They like trout and eel heads, for instance, and will eat them head-first sometimes leaving nothing but the tail. With carp they tend to eat the neck and chest area and leave the rest (much to the disgust of carp anglers). They eat the tails of crayfish but leave the heads and claws. They love frogs' legs. If you find parts of frogs scattered around a pond in early spring this will almost certainly be otters who love to feast as the frogs get together to breed. I once found the head and forearms of a toad that was still alive after being eaten by an otter. Otters will also eat

↓

Otters prefer the tail end of signal crayfish.

waterfowl and seem to favour moorhens. In my experience they don't eat them in any particular way – I have found several moorhen parts after otters had eaten them, including just a foot in one instance. The strangest remains I ever found from an otter was a full-length rat tail but nothing else. They will almost always spraint next to the remains when they have finished eating their meal.

Most Wildlife Trusts in Britain run otter-monitoring schemes and some have training days for volunteers, showing them how to find the signs of otters along rivers. People can volunteer for these schemes and take on specific sections of river to monitor. I thoroughly recommend this to anyone who wants to get out there on the river and learn more about what otters leave behind – you never know, you might even see one!

Charlie: I love filming birds at their nests, especially birds that are difficult to photograph for most of the year. If you have a hide in position you will get cracking views of nesting birds and see interesting behaviour.

Most of the year dippers patrol and hunt along the length of river that makes up their territory. They show up just about anywhere within this area. When they are nesting, however, they spend most of their time hanging around their nest: building it, laying eggs, incubating, hunting, feeding chicks. Putting a hide up- or down-river from the nest means that the birds pass it regularly and often very close.

If it wasn't for the plucky parents all the chicks would have been killed.

I spent the day fiddling around with bits of kit in my workshop. I'm trying to get shots of the dippers underwater to learn more about how they hunt. I've also been to-ing and fro-ing between my workshop and my hide upriver from the dipper's nest all morning, getting the various bits of kit in the water and checking it all works – of course it doesn't.

The light on the bridge is at its best in the afternoon so I'm keen to get back in my hide to try and get some shots of the birds flying in and out. Unfortunately Philippa is off on a shoot and the director has my camera so I'm stuck with the small Sony one. It's a good camera but the lens is not powerful so it's not ideal for filming birds. But nothing ventured, so I head to my hide. When I arrive there I find feathers all over it. I pick up a few and examine them. Not knowing exactly what they are, I pick up a few more. I look around and see that there are lots of feathers. They are dull and brown and have not all 'broken quill', meaning that the feather inside the quill has not fully broken free of it, so the feathers belong to a very young bird. From the way that the feathers are scattered around the roof of the hide it seems to me that a bird has been plucked from the roof by something. I hunt about, scanning the river and bank for a dead bird to try and identify it. In the undergrowth just a few feet from the hide I find the body of a young dipper. My first reaction is to get cross – I'm trying to film dippers underwater and something is clearly trying to kill them. As a wildlife cameraman you can become quite hardened to the death of the animals you are filming and sometimes their death is an irritation if it happened when you aren't filming. It sounds harsh but our job is to document the dramatic lives of animals and a predation by another animal is about as dramatic as it gets.

My feelings of anger subside after a while. When you spend a lot of time with an animal you get to know it a little and you generally get to like it, especially if you have watched it working hard and struggling to achieve something. I am very

fond of dippers, I have watched them all season and I am really gunning for them to succeed at raising a brood. So, as I hold the dead chick in my hands, I realise that actually I do care about its plight.

The dipper chick is half-eaten and whatever killed it was obviously scared off by my arrival. My hunch is that the

The calls are long, high-pitched whistles and they warn other birds.

Just a couple of hours out of the nest – dipper chicks are perfectly camouflaged.

culprit is a jay. When I'd walked down to the hide I'd seen one fly off downriver under the bridge where the dippers have their nest. The signs also suggest a 'corvid' (member of the crow family). Carrion crows, ravens, jackdaws, magpies and jays are all serious predators to nesting birds during the spring and early summer. Magpies have the worst reputation but jays are no darlings. All the corvids are clever – as far as birds go – and this intelligence and cunning helps them work out where birds have their nests and when is best to sneak in to raid them. It also means that once they have found a food source, they don't forget it.

Aware of this I place the dipper chick back where I found it, set my camera on its tripod in my hide and get in. Within minutes I start to hear the alarm calls of robins and wrens. The calls are long, high-pitched whistles and they warn other birds that a predator is near. Some birds vary their calls depending on what they have spotted; this high-pitched whistle means that a bird predator is around. This type of call is known as a 'ventrilocal whistle' meaning that it is very difficult to locate its source. Aerial predators such as

sparrowhawks can hear it but they don't necessarily know where it is coming from. The chatter of the jay, however, is much louder and bolder and as it lands in the tree above my hide I crane my neck to try and get a view of it. The adult dippers begin frantically alarm calling too, but theirs is more of a classic dipper 'click'. The male bird lands on a rock below the jay, right by my hide, and calls repeatedly but to no avail.

I watch as the jay hops down out of the branches, picks the dipper chick up and flies across the river with it. The adult dipper flies up after it but the jay is not put off. It lands in a thicket just upriver from my hide and proceeds to rip the dipper chick apart and eat it.

Of course I'm not managing to film any of this as always my camera is pointing out of the front of the hide. There is method in my madness, though. I am trying to tell the story of the dippers and I know that sooner or later the jay will try and get another dipper chick out of the nest, so I'm sticking with this view of it. I also know that at this time of year jays work as a pair.

It's not long before a second jay appears. It chatters to the other one from a tree behind my hide. I hit Record just in time to film it as it flies downriver past my hide and lands in a small hazel tree opposite the dippers' nest. It sits and watches for a moment – the parents have gone off hunting. I watch nervously, twitching my zoom until I'm happy with the shot – a mid wide-angle of the bridge. I need to get a shot that puts the jay in the same location as the dipper nest and as the bridge is so recognisable I need the jay to fly under its arch to the nest. It does. Hopping off the hazel, it flies straight over to the nest. It is in a small hole in the underside of the bridge's brickwork so the jay has to hang on tight with its claws. It pecks and pokes into the nest and I crash zoom in to try and get my close-up. I spin the lens ring for focus expecting the jay to grab one of the chicks at any moment but it doesn't. Instead it hops off after thirty or so seconds and lands in a tree on the other side of the bridge. All I can think is that the dipper chicks are also smart and have managed to position themselves just out of the jay's reach. The jay looks like he's going to have another go. I hear the click of one of the dipper parents coming upriver, just in time.

The jay launches back towards the nest but is caught mid-flight as the adult dipper spots it and goes for it. The jay veers off back under the bridge towards me with the plucky little dipper in hot pursuit.

The dippers don't leave the nest unattended again after the jay raid, instead they take turns hanging around it. They also become very aggressive towards any other bird getting too close, flying at them and attacking them if they dare land near the nest. One bird, however, is strictly off limits – the sparrowhawk.

It's dawn and I'm in my hide just in time, one of the baby dippers has just left the nest. It is well-camouflaged against the dull brown rocks and water but I can make it out as it nervously hops about, 'dipping' near the nest. It's a Sunday but I had a hunch that the dipper chicks would fledge today so I'm putting in the hours to try and film them – it has paid off. I film the young dipper for a while; getting some safety shots in case I don't see it again. Its parents are feeding the birds that are still in the nest and I get shots of the fledged chick calling to them frantically in the hope that they will come and feed it. A moment later another chick hops out of the nest. It lands in the water below and flaps towards the footings of the bridge. When it reaches them it scrambles for a hold but can't get one, instead it hops through the water until it reaches the small gravel beach where its sibling is. It makes its way towards the other fledgling and the two birds dip together side by side. (Dippers do a lot of dipping, I'm not really sure why. Fred suggests it's because they *like* dipping!)

The two young birds begin to work their way towards my hide. I wonder where they are going but they seem quite agile on the slippery rocks, considering this is the first time they have ever had to walk. Their frantic calls to their parents have

↓
The dippers seemed to be intrigued by their reflection in my lens.

Dipper chicks wait along the edges of the river for their parents to feed them.

paid off and now they are getting fed. I watch in amazement as the female parent pokes around in the rocks next to one of the chicks and pulls out a small fish which I identify as a bullhead. She bashes the head of the poor fish against a rock and then repeatedly pecks at it until it stops moving. She then meticulously washes it in the river, all the while being screamed at by the fledglings, before she hops over to one of them and stuffs it in its mouth. This is the first time I have ever seen dippers catching and eating fish.

The two fledglings continue slowly creeping upriver until they have gone right past my hide and a good twenty metres further. As they are now behind me I can't film them so I turn my attention back to the nest in the hope of getting a shot of the third and final dipper chick leaving it. I wait and wait but nothing happens.

Then something lands right on top of my hide. I freeze and peer up slowly, trying to work out what it is. I can just make out the silhouette of a bird but can't tell which species. There is no way of getting a look as it is on the other side of the canvas roof and only three inches from my face – any slight movement would spook it. It sits

motionless for at least three minutes then suddenly flaps off. There is an eruption of bird alarm calls as I peer out of my hide window to get a look; a second later a sparrowhawk flaps right past me with something in its talons. It lands on the bridge above the dippers' nest. The adult dippers go wild beneath it. I tilt the camera up at the sparrowhawk, hit Record and zoom the lens in. It has caught one of the dipper chicks. The small brown bird lies lifeless in its talons. The sparrowhawk hops off the bridge and flies downriver. The adult dippers chase after it but keep their distance – no small bird messes with a sparrowhawk.

To my knowledge the dippers did successfully rear two of the four chicks they started with and spent a couple of weeks feeding them below the weir outside the house. As much as I love dippers, the relentless sound of the chicks begging for food drives me mad. Not only is it so constant but it sounds very similar to a kingfisher and so I am constantly being distracted by it. The dippers also went on to successfully raise another brood of chicks under the bridge – so, despite the casualties, a good year for them.

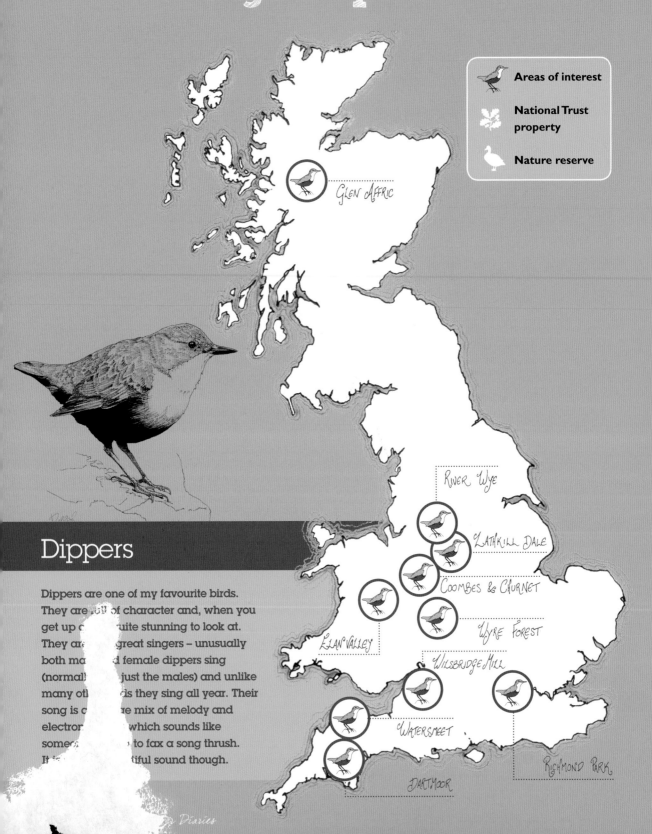

Areas of interest

National Trust property

Nature reserve

GLEN AFFRIC

RIVER WYE

LATHKILL DALE

COOMBES & CHURNET

WYRE FOREST

LLAN VALLEY

WILSBRIDGE HILL

WATERSMEET

RICHMOND PARK

DARTMOOR

Dippers

Dippers are one of my favourite birds. They are full of character and, when you get up close, quite stunning to look at. They are also great singers – unusually both male and female dippers sing (normally it's just the males) and unlike many other birds they sing all year. Their song is a strange mix of melody and electronics which sounds like someone trying to fax a song thrush. It is a beautiful sound though.

Glen Affric, Scotland

If you want to see dippers in one of Britain's most beautiful locations try Glen Affric. The glen itself stretches from Kintail to Strathglass, thirty miles east. The area is bisected by a number of burns which all eventually merge to form the River Affric. All of these watercourses are fast, rocky and well-oxygenated. Follow the River Affric walk for easy access to the river.

River Wye, Monsal Dale, Peak District

Park at Taddington Wood, cross the road at Lee's Bottom and start a walk through Monsal Dale. The stunning little river is fresh and powerful and supports a very healthy population of dippers.

Lathkill Dale, Derbyshire

Starting as a trickle at Fen Dale, the River Lathkill winds its way through the stunningly beautiful Lathkill Dale. This is perfect dipper habitat and as a result supports a good population. There are plenty of paths in the dale so stick to them, listen out for the call of the dipper and watch as they feed and preen in close proximity.

Coombes & Churnet, Near Stoke on Trent

The habitat at Coombes Valley nature reserve ranges from hay meadows to woodland. The Coombes Brook sits at the bottom of a deep valley and is a magnet for dippers. There are no food and drink facilities at the reserve and the nearest shop is four miles away, so go prepared.

Llan Valley, Wales

This area of mountains and reservoirs in central Wales is definitely worth a visit – it is so stunning. It is made up of five reservoirs and is known as the Welsh Lake District. Dippers can be seen along the rivers that flow out and into the reservoirs which eventually empty into the upper Wye. Access is easy due to numerous roads and tracks.

Wyre Forest, Worcestershire

Straddling the Shropshire–Worcestershire border, the Wyre Forest is 6,000 acres of stunning native woodland and one of the largest areas of ancient woodland left in Britain. Running through it is the Dowles Brook, a good-sized, rocky river – perfect for dippers. The birds are present year round and can be seen from various spots. Parking is available at various sites including the Forestry Commission Visitor Centre, three miles from Bewdley.

Wilsbridge Mill, Hanham, Bristol

Wilsbridge Mill is run by the Avon Wildlife Trust. The Syston Brook is a smallish stream that runs through the mill before it heads down towards the River Avon. The stream is perfect lowland dipper habitat and seeing them there is pretty easy. They're used to people so are quite approachable.

Richmond Park, London

London is not good for dippers. However they can be found if you know where to look. The Beverly Brook that flows through Richmond Park occasionally turns up a dipper and is definitely worth a visit. Access is great as there are several car parks near the brook and walks along it – good luck!

Watersmeet, Near Lynton, North Devon

Watersmeet is one of the most picturesque spots in North Devon. The steep-sided hills and wooded combes make up some of the largest areas of ancient woodland in the south west. There are numerous streams and rivers in the area to explore. Watersmeet itself is the spot where the East Lyn River meets Hoar Oak Water, forming one of Britain's deepest river gorges. There is an Edwardian t room and garden there – a perfect place to sit and wa   rs!

Dartmoor, Devon

Dartmoor is a vast area of wild rugged moo   uated in the heart of Devon. The rivers Teign, Dart, T   and Taw all start here and are great for dippers. Dar   riss-crossed by numerous footpaths and tracks   asy.

Philippa: I like to think that I am aware of the species we have around us, and this year of studying them so intensely is going to bring us close to them, but it is only now I concentrate on which creatures we will film that I begin to think about the species we no longer have on the river.

There is one obvious one, the beaver. It was once native to Britain but had been hunted to extinction by the seventeenth century. I often wonder what it would be like to look out of the kitchen window and watch beavers swimming by. There is something comical about the thought. Would there be any trees left? Would the river be blocked up with dams?

But now I come to think of it, there are other creatures missing. Ones that are less obvious. When was the last time we saw an eel or a water vole? What is really going on along and under this river? We have become so taken up with what we can see directly outside the house that we simply forget what we can't.

What of the water vole? A small rodent, seemingly insignificant and very often mistaken for a rat, the water vole of our rivers is indeed 'Ratty' from our childhood – a swashbuckling, oar wielding, adventuring picnicker.

The vole is not at all like a pointy-nosed scaly-tailed rat; think of the chubby-cheeked hamster family pet and paint it brown, give it a tail and you aren't too far off. And water voles are vegetarian, rats are not.

The water voles on our river are the largest members of the vole family: they look cute and they have brown fur which continues on the tail and covers their ears, so that they appear not to have any. They live in tunnel networks in river banks and are well adapted to life in and out of the water, keeping warm by trapping air bubbles beneath their fur so that their body doesn't get wet. They are eaten by stoats, herons, owls, brown rats and pike; an otter might take one occasionally, but the most significant predator is the mink. They only live for about three years but in all likelihood only last about five months in the wild.

When we start to ask around, the older neighbours remember them being commonplace; the farmer upriver says 'they were all over the place' when he was a lad. Charlie remembers being twelve in the mid-eighties and watching them plopping into the water and swimming around from a bridge further upriver from where we now live. I ask him what they were like but he says he didn't bother to take much notice of them because they were always there. Yet now we don't see them at

The water vole is the fastest-declining mammal in the UK.

all, they are certainly not part of the river life my children are growing up with. Why not?

The water vole is the fastest-declining mammal in the UK principally because of mink. This voracious predator was introduced in the 1920s for fur farms, some escaped and others were released by well-meaning but misguided animal rights activists. They very quickly naturalised and started to eat water voles for whom there was no escape. Over many thousands of years water voles have co-evolved with their natural predators and they avoid them by creating small tunnels with the main entrance under the water. The only predator that could squeeze into the tunnels is the stoat but it is not adapted to the water. The mink, however, was never part of this co-evolution process and so the water vole has no defence mechanism in place for an animal that can and will swim underwater and into the entrance tunnel where a whole colony of voles is easy pickings.

Long-term there is hope, thanks to the return in recent years of the otter. The otter is far bigger and more powerful than the mink and it looks as though it will out-compete it. We have certainly seen fewer mink on the river in recent years and more and more otters. When we first moved here we saw mink regularly. They are very confident and usually easy to spot, they would swim past the house with their heads up and once I glanced up from my desk to see one peering into the office door. Not even when our eyes met did it move until *it* was ready.

Now I can't remember the last time I saw one. There is much rumour amongst those who spend their time on rivers observing these creatures, about the return of the otter and the demise of the mink; there are fascinating tales of mink and otter in face-to-face combat, otter incisors even being found in mink skulls. But we can't be sure how often otters beat up mink, whether this is happening regularly and what impact it has on the overall population. However 'out-compete' mainly refers to the fact that the otters are simply eating most of the food available to the mink.

Even if mink numbers are greatly reduced they can't have completely disappeared. Is it too late for the voles here? Have they all been eaten? We have been here ten years and in that time I don't remember ever seeing a water vole.

Since 6th April 2008 the water vole has been protected. But what use is that here if we don't have them on the river in the first place? The wildlife trusts play a valuable role in water vole reintroduction and it turns out that it isn't just our river which is missing an integral character, they have disappeared from almost 90 per cent of the sites they occupied sixty years ago. A common every-day species just slipping away under our noses.

And it's not all because of the mink. Mis-management of bankside habitat and accidental poisoning are also to blame. But there are many things that we can do to help. As I read about 'Ratty', far from getting depressed I find myself getting excited. This year could be the year that we bring the vole back, the year that we make a big difference. We have the chance to do some practical hands-on conservation that will provide a positive legacy for generations to come and benefit the wildlife on this wonderful river.

First of all I need to find out whether there really are any voles left here, if not then we have nothing to help. My first call is to the local wildlife trust.

Despite their fearsome reputation, mink are actually quite cute.

Charlie: I'm sitting in my hide at dawn. Richard (director) and Stephen (soundman) are both sitting in a hide about fifty feet behind me. My view is downriver to the nest the kingfishers have been working on for a few weeks.

I'm in the hide this morning as I want to film the kingfishers mating and I reckon today is the day. It's cold though and I'm bored. I usually try to drag out my snacks and coffee or play solitaire on my phone. But this morning I'm not playing games, I can't afford to miss the action. Filming kingfishers mating is very difficult and the only way to really guarantee getting it is to spend a lot of time in the hide waiting for them to do it. Still, being bored makes me distracted. I phone Richard in the hide behind me to moan about something. Richard as usual placates my moaning and listens to my ramblings. Of course while I'm on the phone I'm watching the kingfishers. The female is in the nest digging and the male is sitting outside. I watch the female fly out of the nest and land on the perch outside the bank while I witter away. The male begins making a very distinctive whistle, a series of repetitive single peeps; I suddenly realise what I'm listening to. 'They're going to mate!' I whisper down the phone to Richard and hang up. The male flies onto the female, I hit Record and spin the camera on to the birds and zoom in, hitting Record again because I'm panicking. Now pressing Record twice is never a good thing to do. One command puts the camera into Record mode, the other turns it off. I get my focus and watch with relief as they're still mating. Then I realise there is no red light on in my viewfinder, I check the display on the side of the camera and nothing is happening. I hit Record again and the camera begins recording just as the birds split up – I have missed it! It's OK, though,

I have an eight-second pre-record function on my camera which captures the eight seconds before I hit the record button. I always leave it on for this very eventuality; except today when it seems to be off …

I phone Richard and explain that I missed it. I won't repeat what his reply is. I assure him that it's all OK, they'll do it again in a minute. Half an hour later I'm bored so I phone Richard again. I keep my eye on the birds and exactly the same thing happens. The female flies out of the nest and the male starts whistling. This time I'm ready, I drop the phone after the first whistle, hit Record, zoom in, focus on the female just in time for the male to fly up, hover over her and land on top. They mate for about twenty seconds before splitting up. The male then flies off down river, excited and whistling, leaving the female to compose herself. Got it! I am hugely relieved. Not only have I redeemed myself but I have got a key moment in the life cycle of the kingfisher.

I leave the birds alone for a couple of weeks after this. I have a huge list of other sequences to shoot and although I could film nothing but kingfishers for ever, I do have to tear myself away occasionally. This time of the year is very quiet for kingfishers, so I won't be missing much. They will continue to mate several times a day for a week or so before the female lays her eggs. She will lay one a day for about a week before she starts incubating them. This ensures that they will all hatch together; which will be about three weeks later. So the nest area will stay pretty quiet with just occasional

visits by the birds as they swap over incubating duties.

Three weeks later, I'm back at the kingfishers' nest in my hide. I've been sitting for two hours and I've not heard nor seen anything. I am beginning to get worried. If the chicks had hatched I would expect to see an adult bird going in every half hour or so with a fish to feed them. If the chicks hadn't hatched I would expect to see the adult birds changing over incubation duties every hour and a half, or so. This morning I have seen nothing. I decide to give it another hour.

An hour later still nothing. Something is wrong. This nest is not working. I leave the hide, grab my torch from my bag and go to take a look up into the nest. Kingfisher nests are extremely sensitive and looking up them with a torch is not at all advisable. I have spent my life working with kingfishers, though, and I know that there is something wrong with this nest. My torch immediately reveals the problem. Halfway up the nest tunnel is

a broken egg. This nest has been raided. I scan the inside with the torch, trying to get a look into the chamber. If the adult bird had been in the chamber when it was raided it would have died and I would have expected to see blue feathers. Fortunately I don't, although this doesn't mean much. I inspect the bank below the nest and can see scratch marks from claws. I can see some around the tunnel entrance too, most probably a rat. Rats are a real problem for a huge number of nesting birds and one of the kingfishers' worst enemies. Kingfisher nests are specifically designed to keep intruders out. Kingfishers select banks that are high and are either inclined or vertical – never sloped. This means that it is very difficult for predators to climb up the bank and get to the nest. This bank looks to me like the perfect location so I'm amazed that a rat has managed to get in. I have seen, and indeed filmed, raids on kingfisher nests by predators and they are always in nests that the kingfishers have chosen badly.

Rats are a real problem for a huge number of nesting birds and one of the kingfishers' worst enemies.

I'm gutted. I really love kingfishers and when they die like this or nests fail, it makes me very upset. What I really want to know, though, is whether or not the adults have survived.

Luckily I know all the good nest banks on the river so I set about looking to see If I can find the parents and whether they've nested again. I check upriver in the woods; there are two banks up there where kingfishers often nest, although there is a territorial border on the edge of the woods and I would be surprised if the pair had crossed it. Further downriver is

a very high steep bank which I call the ochre or whirlpool bank. It is dark red because it is made of ochre and the pool beneath it is the deepest part of the river. There is a very large overhang making it the perfect place for kingfishers, as it is totally impenetrable. I hate it when the kingfishers nest in this bank as it completely stifles my work – I can't see what's going on!

I choose a spot on the bank opposite, about thirty metres away from the nest set in the undergrowth, well-hidden from the kingfishers. I set my camera up, get

A small wooded river is the perfect environment for kingfishers

into the hide and wait. About an hour later I hear a whistle. I peer through the hide window trying to see where it is coming from and catch a glimpse of a kingfisher as it lands in the roots of a tree by the bank. It's the female. I swing the camera on to her and hit Record. She whistles a couple of times. Out of the corner of my eye I notice the flash of another bird. It must be the male coming out of the nest. The female then whistles a couple more times and flies up under the overhang and into the nest. Found it!

I am really chuffed. Not only have both birds survived the nest raid but they've quickly got themselves back on track with a new nest. This means that in a few weeks the river will go crazy with kingfishers again and I will be a very happy man!

A few weeks later

Spring has sprung and gone and today is truly a summer day. I'm back in my hide opposite the whirlpool bank where the kingfishers have their nest. Yesterday I saw a kingfisher fly past the house heading upriver with a fish in its mouth; this means that it was off to feed another bird. My suspicions are that it was the adult male feeding the chicks. The only way to confirm this, of course, is to sit quietly in the hide and watch to see if an adult appears with a fish and takes it into the nest.

The wait is longer than I expected, nearly an hour. I don't see the kingfisher at first, it has appeared unannounced on the lower branch of a short willow tree. It watches my hide for a while, checking the lens out. I freeze and allow it to settle. It does. I hit Record. Looking through the lens I can see that it's the male and it has a very large juicy minnow in its beak. It holds the fish with its head facing

outwards; this means that it can feed it to the chicks head first. Swallowing a fish tail first can cause the kingfisher serious problems. The kingfisher whistles once then flies into the nest. I count him to see how long he takes. One, two, three, four, five, six, seven and he's out. This means that the chicks are more than a few days old. Kingfishers are born blind and helpless and it takes over a week before they're in a position to see and accept food from their parents. It is dark in the nest and the only way they know food is on offer is when it is presented to them. As they get to about fifteen-days-old, they can see and are more robust, so they start to react to the light changing as the adult bird enters the nest chamber and blocks the light off. I have also noticed from watching kingfishers in the nest that the parents make a short rasping noise as they move up the tunnel towards the chamber to let the chicks know they have food. So seven seconds in the nest means the chicks are fairly well developed. Also, as the male exited he belly-flopped straight into the river before flying off. Kingfishers do this to wash after being inside the nest. As the chicks get older the nest chamber becomes truly disgusting as it is full of fish bones and excrement. The parent birds, being proud and clean, don't like this so they wash as they exit. However, they only do this once the nest is really filthy, usually a couple of weeks after the chicks have hatched. These clues are essential so that I can get my timings and schedule right. I have a busy year filming and I need to know roughly when the peak kingfisher activity is going to occur.

This is a very busy time for the kingfishers. Not only do they have to work incredibly hard feeding their six or seven chicks but they have to start digging a new nest and courting each other ready for their second brood.

Philippa: **Wild garlic grows everywhere along the river at this time of year, the smell is along every bank. The leaves begin to poke through the damp, bare, brown earth by late February and I have got to the point now where I look out for them.**

Wild garlic is the first great harvest of the year. The plants continue to grow on through March at which point, when they are still small and fresh, they are really at their best. It is said that bears forage for wild garlic for its stomach-cleansing properties. Well, although I do have the odd grizzly moment, I'm no bear.

There is one tree-covered bank I visited today and, underneath those trees is a swathe of wild garlic. Now it is in full bloom and the star-shaped flowers take your breath away, seen en masse or even as individual tiny shapes. They have a delicate beauty that matches the delicate flavour of the leaves, and yet every year they are overlooked.

The bank I head for is steep – it is just a short slide down to the cold water. I crush the wild garlic with every step and the wet leaves don't make the safest of paths but all I want to do is get to the middle. There I crouch down, breathe in and just admire the flowers; the sun is low in the sky and soon it will be lost behind the bank altogether but for now it manages to back-light much of the top section of the swathe. The garlic leaves are luminous fresh green and the dancing white flowers on their thin stems shine out. A moorhen calls. It has only just realised I am here and she runs along the water below making such a fuss that if I were a predator I could snap her up. I pick just one flower stem and hold it up to the light. Moments like this that come just once a year are meant to be savoured.

This is not a plant to be taken for granted, I love to use it and yet it is so ephemeral. At the end of each spring as the season ends I am filled with regret for the things I didn't have time to try.

I won't be scrabbling around for bulbs today; it is the leaves I am after, and they are only with us for such a short time. Just like spinach or rocket these leaves are hard to keep and so it's best to pick them and use them immediately. There are

Moments like this that come just once a year are meant to be savoured.

ways of freezing them in a dish – they are great in a flan or quiche, for example, but they are by far best picked straight from the plant. Every year I find a new way to use wild garlic.

Wild garlic, like its culinary cousin, is nutritional gold, but only recently are studies revealing just what a health enhancer this unassuming plant is (see above).

And here it is growing on the riverbank for free. What's not to love?

With my basket of leaves I head straight home for wild garlic on bread.

Wild garlic pesto

Use the leaves not the bulb and, as with any foraging, be careful. If you are nervous of using wild plants in cooking then this recipe is good to start with because it's so easy to identify wild garlic from the smell; just the same as regular garlic. Try to use the smaller fresher leaves and ideally before the plant is in flower.

Wild garlic leaves | **pine nuts** | **olive oil** | **Maldon sea salt** | **grated Parmesan**

Roughly chop the wild garlic leaves – don't do this in a food processor because the leaves are so wet the whole thing tends to turn to mush. After some experimentation I have discovered that one of those curved knives used for chopping herbs is ideal.

Lightly toast some chopped pine nuts, leave these quite large – it's lovely to have their crunch.

Mix the garlic and pine nuts with some olive oil – a bland one that's not too peppery – and salt.

Toss into a bowl of warm pasta and top off with Parmesan.

Save any extra in a jar in the fridge for up to two days – it is great on bread for a healthy lunch or stirred into rice, and you can still do the school run afterwards without people fainting in your wake.

The wild garlic breath test

You know you love someone if you are prepared to try this. Eat a large quantity of wild garlic, chopped and mixed with butter and then spread onto warmed bread. Then when you have a volunteer, breathe heavily on them and see what the reaction is. Can you smell it? We can't.

Filming dippers underwater

Charlie: Ever since I started spending time on the river I've wanted to try and get shots of dippers underwater. I've come up with various ideas on how to achieve this but have never succeeded. It was only when I was looking out of the kitchen one day watching the birds on the feeder that I came up with the answer – an underwater bird table.

The theory was this. If I could get a dipper locked on to a feeding station then I could move the station underwater and hopefully film it. My first attempts failed. I bought some meal worms from a bird food supplier, stuffed them into a specially made net envelope and placed it underwater in a channel that the dippers seemed to be regularly hunting in. After placing it I snuck back to my hide and waited to see what happened. It wasn't long until the dipper turned up. The dippers had chicks downriver in a nest underneath a stone bridge so they were spending all of their time foraging, which meant they were always around the same spots. I watched with bated breath as a dipper began hunting up the small channel where I'd placed the bait.

Dippers hunt by turning stones over on the river bed and grabbing the insects as they scuttle off to hide. They are meticulous when working an area and leave no stone unturned. So, as the dipper approached my net envelope I was sure it would find the tasty meal worms. I zoomed my camera in and started recording as the dipper moved on to it. It dived down and then a second later popped up with a meal worm. Victory!

● REC

The dipper very quickly locked on to the free food supply.

bit of kit I made a few years ago for photographing kingfishers underwater. It is essentially a cube of perspex about twelve inches long on each edge, with a see-through dome port on one side. Dome ports are used by underwater photographers to shoot pictures – they correct the distortion created underwater and give a slightly clearer shot. The cube has no top so it can't be fully submerged. Once it is in the water a camera can be placed inside with the lens pointing through the dome port. The camera can be fiddled with from above and then wired back to the hide. The biggest problem with the cube, though, is that it displaces so much water and also floats – so I have to use lead diving weights to keep it on the bottom.

I placed the cube in the water right next to the dipper's rock, put the camera in and then set the ramekin full of insects back on the rock – I wanted to get the bird used to the cube and wires before I tried to get it to go underwater. The dipper turned up as I withdrew to the hide and immediately sat on the cube to inspect it. The cube seemed very interesting to him and he went inside, poked about, pecked at all the wires and then started to attack it! Being made of shiny black perspex the cube was basically like a mirror to the dipper – a bird which is highly territorial. So rather than seeing a rather strangely constructed Heath Robinsonesque piece of filming equipment it saw itself – an aggressive intruder. The dipper then became completely preoccupied with its own reflection and spent its time vandalising it. In the end I had to cover every shiny part of the cube that I could with gaffer tape and water weed until the dipper stopped looking for itself and started eating out of the ramekin again.

A couple of good sessions later and I decided it was time to go sub-aqua!

And then almost immediately tossed it away. A mistake, I thought, he'd go back down for more, but he didn't. It seemed that dippers don't like meal worms.

My next plan was to use the insects that the dippers were hunting for. These include lots of tiny underwater insects – crustaceans called gamarus, cadis fly larvae, beetle larvae, mayfly larvae; anything I could catch by turning stones over. I put all the insects I'd collected into a small white ramekin and placed it on a prominent stone in the middle of the river. One way to woo a dipper is to give it a big stone in the middle of the river – they love 'em!

To my amazement the dipper found the food in the ramekin within ten minutes of me getting in the hide. The stone was so irresistible that the dipper landed on it straight away and the moment it saw the insect larvae inside, it devoured them. The plan was working.

I spent a day getting the dipper eating out of the dish until it was so used to it that it would just go between the nest and the ramekin when feeding its chicks. It meant I spent lots of time catching small insects, but it was working.

The next day I went down to the river armed with 'the cube'. The cube is a

↓
'The cube' – it may not look like much but it can get great results.

So how do you keep a load of underwater wrigglers in a submerged ramekin? I thought about this for a long time! Then it occurred to me – clingfilm. I put the insects in the dish, filled it with water, then covered it with clingfilm. A rubberband wrapped around the ramekin kept the clingfilm in place and I was ready.

I placed the ramekin into the water in front of the cube and surrounded it with stones. The water was only seven or eight inches deep so I was hoping that the dipper would be able to see the food inside from its rock. I fired the camera up inside the cube and retreated with my remote-recording button – this allowed me to put the camera into record mode.

The dipper was back on the cube immediately and hopped about looking for its ramekin. It wasn't long before he spotted it. I put the camera into Record and waited nervously. The dipper was apprehensive. He stood on the edge of the dome and peered in. He then thought better of it and flew away.

This is classic wildlife film-making. You have a plan, you put the plan into action with total confidence, the plan fails. But if you persist, you often succeed. So slightly irritated, but not undaunted, I continued.

The dipper came back several times and did the same thing, as well as occasionally finding gaps in the gaffer tape and attacking itself. In the end, though, he just couldn't resist and he went for it.

The shot only lasted a few seconds. The dipper paddled around above the ramekin, looking in, then flapped down and stabbed at the clingfilm. Breaking through it he grabbed an insect and was away. Success! The shot was in the bag. I tried again a few times to get better pictures but the dipper never did it again. The image is, to my knowledge, the first-ever shot of a wild dipper underwater.

Charlie: Moorhens are much overlooked birds, in my view. There are several that live outside our house and Philippa and I are very fond of them. Moorhens that live in water frequented by humans become tame so can be watched fairly easily.

Our moorhens are not tame, in fact the merest sight of a human sends them scuttling for a reedbed or bit of undergrowth to hide in. It is watching this behaviour that has made me a great admirer of the species – when you see them hiding and sneaking you realise just how smart they are when it comes to keeping hidden.

This orange part was the only thing sticking above the water. I pointed it out to Bill and he put his binoculars on it whilst I waded into the water for a closer look. 'It's a moorhen,' said Bill. I stopped and looked hard. He was right, a closer look revealed the contorted shape of a moorhen under the water clinging to a root. Its head was up and its beak was just breaking the

They are up there with otters and leopards – able to outwit us and other predators.

I was once doing some filming on the river with veteran birder Bill Oddie. I was supposed to be filming Bill watching kingfishers and there was a crew of around seven of us. We were looking for a location to shoot the piece and were all standing on a small beach on the edge of the river (about a mile upriver from the house). I kept noticing something in the water but couldn't work out what it was. It looked like a piece of rubbish, a weird sort of shape like a crumpled crisp packet, silver in colour with a bright orange bit at the top.

surface, allowing it to breathe. I withdrew and we filmed a piece about it – which never made it to air. Both Bill and I were amazed by the sneaky moorhen, though.

I have had similar sightings on the river since. Often the birds will flee as they see me coming up the river – I even watched one swim underwater for over thirty metres to avoid being spotted! But if I surprise them they hide instead. Often they are so secure in their ability to hide that I can almost stand on them without them moving. This tactic of not moving but relying on hiding to avoid danger is not uncommon among animals, but most do immediately flee when they see a human. In my experience those that hide instead of running are those most confident in their abilities – otters are the best example. Otters know that we can't see them if they don't want us to and as a result they will happily swim right beneath our feet and we will never know of their presence – I have filmed this

Building a nest requires time and skill – and the perfect twig.

happening to a fisherman. Leopards are the same, able to blend and vanish into their habitat so well that if they don't want us to find them, we won't – even when they are right next to us. So moorhens to me are not just boring little 'water chickens' as my neighbour calls them; they are up there with otters and leopards – able to outwit us and other predators whenever they choose.

This doesn't mean that they are clever, of course, it just means they are masters at vanishing. While filming *Halcyon River Diaries* I concentrated a lot of my time filming moorhens and one particular episode proved to me that these birds are no Einsteins!

April/May

I'm armed with a box of kit – two cameras, infra-red lights, mounting arms for the cameras, and lights and wires of all lengths and uses. I've got all this at the bottom of the moorhen tree. About fifteen feet above my head is a moorhen nest hidden amongst the ivy that has wrapped itself around a hazel tree. I put a few bits in a small rucksack, tie some wires to my belt and head up the tree to the nest. The tree is not large (about thirty feet) and not very strong, so the climb up is precarious and difficult. After a couple of minutes of heaving, stretching and moaning I'm close to the top, the only place I seem to be able to get a view into the moorhen's nest. It is very neat, nestled in the crook of a branch, with six pale eggs inside it.

Time is of the essence. Birds' eggs will chill and die if they are left unattended for too long. I get rigging and within seven or eight minutes have the various bits set up in the tree. I rainproof everything and

REC

Moorhens don't usually nest in trees – and in my experience shouldn't.

shimmy down. The moorhen is back almost immediately and as I sneak off with the wires she flies back up into the branches.

I run the wires back to a ducting pipe that I have in the garden and attach them to the ones that stick out of the pipe. These wires go directly underground through the garden, across the bridge, under the patio and through the wall of the house into the cupboard in the kitchen. I put them in years ago when we were having the house done up and they allow me to watch the life in the river from the warmth of the house – I can even patch them around the house allowing me to watch the remote cameras on the living-room television or the one in our bedroom. This time I'm working by the sink. I plug my monitors into the wires coming out of the wall and an image of a moorhen on her nest appears. It's not the best quality in the world, but it will do.

Having live pictures of animals in the house means that the kids get to watch their lives at close range. They have become very blasé about this; we used to have the otter holt wired live to the television in the bedroom and the kids would lie in bed and watch live otters every night. They are keen on the

moorhens, though, which pleases me, especially as the moorhens don't actually do anything for nearly two weeks other than sit there looking about while keeping their eggs warm.

One morning I come down to make Philippa a cup of tea and I switch the monitors on. One of the moorhens is on the nest and as I watch it stands up, revealing three chicks! I call the family down excitedly and we watch the chicks struggling about in the nest. They are small black bundles of fluff with clumsy oversized black feet and weird bald red and purple heads. They are strangely appealing and the kids love them, watching in amazement as they stumble about the nest. It is moments like this that make the huge amount of effort rigging cameras along the river worthwhile. The kids are often not remotely interested in some of the wildlife spectacles I show them, but on other occasions they become completely absorbed and watch with great excitement.

Moorhens fledge very soon after hatching. These ones in the tree are an uncertainty to me, though. When moorhens nest in reed beds the water is just a small hop away and the chicks will be taken on outings a day or so after hatching, climbing back into the nest

before they get too tired. Nesting in a tree means that going out on small excursions is out of the question – once you're out of the nest there's no going back in.

I watched the moorhens on the monitors closely for a few days, waiting for the moment to make my move and get into the hide near the nest to film the chicks fledging. Filming chicks just after they leave the nest is crucial to any story about them – it is the moment when they are at their most vulnerable and therefore a key sequence to get when telling their story.

The moorhens hatched four days ago and I'm sitting in my hide filming the last dying breaths of the first chick to fledge. It is still moving its head trying to keep it above the water but it is basically doomed. The parents are swimming around it. They come over and inspect it every minute or so before continuing to search for food to feed the other chicks. Some instinct tells me to wade in and rescue it but I'm not going to. Firstly, if I was going to do that I should have done so already because the chick is past saving, and secondly, it would be a very rare moment when I would intervene to save a chick which is with its parents. If I were to help it I would almost be compromising it more as I would then have to rear it and that would be completely unnatural – the chances of the chick ever surviving back in the wild would be slim. People often ask me if I want to try and save animals that I see in trouble. Often I will but only if the animal is suffering or stands a good chance of being saved by my intervention; I once reared and released an otter cub which would certainly have died without being rescued. Sometimes it works, sometimes it doesn't. Rather than having a black-and-white rule about it I make the decision on each case, and only in Britain – I would never attempt to save any animal on the plains of Africa, for example. There I am just an observer.

I watch the chick raise its head for the last time. It blows a single bubble from its beak and dies. I film it as it floats off down the river and feel a tinge of sadness. I am a bit like a war correspondent in a sense. I see a lot of animals killing each other and dying and get quite immune to it. But I am caught out occasionally especially by animals who I have watched grow up, such as this moorhen.

I'm not sure why the moorhen has died, I assume the jump from the nest into the cold river just shocked it too much. Moorhen chicks die very easily and something as simple as exhaustion or shock will kill them within minutes. I watch as one of the parents climbs back through the branches up to the nest to feed the five remaining chicks.

The next day when I look at the monitors I see there are only four chicks left. I go out to check the river near the nest and see the body of another chick in the water downriver from the nest tree. Over the next three days one chick a day jumps from the nest and dies until finally there is just one left.

I'm watching the monitor closely – the last remaining chick looks like it is about to fledge. I hit Record on the video

Moorhen chicks are very fragile and die quickly from shock or cold.

monitors in the kitchen and leave Philippa watching them while I dash up the garden with my kit to try and film the chick leaving. By the time I reach the hide it has already jumped. I sneak into the hide without the parent birds seeing me and focus my lens on the last of the chicks. It is obscured by a small branch in the water but both parents are with it and trying to feed it, so I have high hopes. I pick up some shots as the parents grab flies and various insect larvae from the bottom of the river. Eventually the chick emerges from behind the log. It doesn't look good. It is still afloat but its head has drooped to one side. One of the parents swims over and tries to feed it but it doesn't respond. The other bird comes over and both attempt to feed it but it is useless. They give up and start preening next to the dying chick. The chick rallies briefly as it attempts to climb out of the water on to a root but it then gives up and dies – it can't have been in the water for more than fifteen minutes and it's dead. One of the adult birds pecks at it a few times and then swims off downriver.

I feel gutted for the moorhens, six weeks of hard work completely down the drain. I always thought it was an odd decision for them to nest in a tree. It turned out to be fatal.

Charlie: OK, let's start by saying that filming otters at night is very, very difficult! But it's not impossible and with a few bits of fairly inexpensive equipment, anyone with the drive and patience can do it.

Firstly you need to find your location and confirm that there is an otter regularly working the area by following the field signs described in 'How to track an otter' on page 34. Once you've established that then you need to sort out the technical bit.

There are many ways of filming at night and all cameramen have preferred methods. Some have very expensive highly-engineered rigs; others, like me, have kit that is essentially held together by sellotape and Blu Tack. I don't really care how the kit looks or how well-engineered everything is, I just want it to work and get my shots. I have refined it a little over the years but it is much the same kit that I was using ten years ago, the only difference now is that it is much cheaper to buy!

The most obvious and basic thing you need to start filming otters at night is a video camera. Now, there are two types – cameras with 'night shot' mode and cameras without. 'Night shot' mode is very useful. It allows you to film at night by changing the filter set-up in the camera, enabling the camera to be more sensitive to low-light and infra-red. The night-shot mode is not, however, sensitive enough to pick up an otter at a distance. A big powerful spotlight or torch can solve this – by mounting it on a tripod next to the camera, you quickly have a fairly cheap and easy rig.

The equipment I use is slightly more complex but still fairly achievable. It consists of a security camera with a zoom lens attached to it, wired into a video camera. The security camera

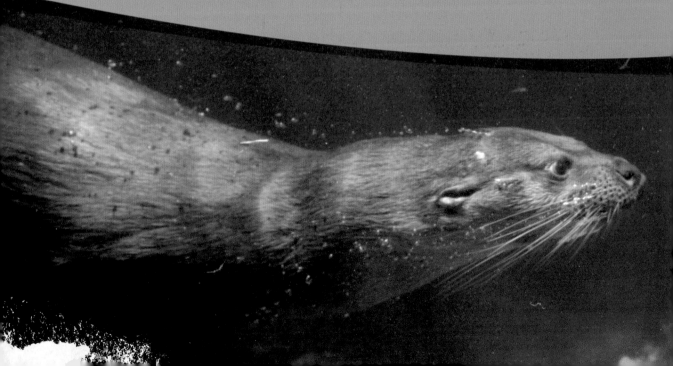

is a basic off-the-shelf model (I use Ikegami 47). These cameras are relatively inexpensive and are highly sensitive to low-light and infra-red. The lens on the front can be anything you like – you can buy second-hand manual focus lenses for next to nothing these days. The important bit is finding a mount to attach the lens to the camera – there are specialist companies that make these. Once you've got a camera and lens you need a recorder. I use an old video camera. It has two settings – cam and VTR. Cam mode means that it just works as a video camera and VTR mode is the mode you switch it into if you want to watch what you've shot. However, VTR mode also works like a video machine and if you have a camera with an A/V in socket you can basically plug anything into it and record on it (just like a video). So with the video camera in VTR mode I can plug in my security camera, see the image that the security camera is seeing and record it.

Obviously everything needs a bit of power to keep it going. I use motorbike batteries in a back pack.

Now all you need is a big torch or spotlight. I use a Clulite hunting light but you can pretty much use anything bright enough. You can filter the torches with infra-red filters, red filters or nothing. As far as I can see, otters don't take any notice of light when you shine it at them. But I tend to use red or infra-red gels to cover my lights. Red is less intrusive when filming at night because a lot of animals can't see it. So it means that not only can I film otters but I can also film anything else I come across.

Red is less intrusive when filming at night because a lot of animals can't see it.

Philippa: **Today we surprise the kids. We pack them into the car on a mystery mission and they try to guess where we are going for the whole journey. I tell them that we have entered a pig-licking contest, so when we open the car door Arthur simply wanders around asking to see the pigs.**

At first you can see why, the one-stop duck shop is in converted stables next to fields of happy chickens, with ponds full of contented ducks. But inside the warm quiet stables is the most exciting part. Each loose box has been adapted with a gate, filled with soft bedding and has a large heat lamp over it. When we peer inside, they are full of chicks. Stripey, speckled guinea fowl babies in one, yellow ducklings in another, brown ducklings in the next and assorted chickens in the next one along.

All erupts into happy chaos as the three boys peer over the gate and realise why they are here. I think, in all honesty, that I am actually the most excited. Before the

all-important moment of choosing the ducklings we need to gather duck essentials. So in the small shop at the end of the row of loose boxes we gather feed trays, a book, food and shavings. Amid all the noise and trying to concentrate on what we really need and what we don't, eventually the sound of Gus repeatedly saying, 'We need this', permeates. I turn around to find that he is staggering under the weight of a large brown paper sack of grain which he is attempting to hoist onto the counter. Luckily for him, but not great for the rest of us, the weight of that bag is rapidly diminishing because he has managed to rip a hole in the bottom of it and the grain is simply spilling out all

over the floor. Arthur is chasing the resident Jack Russell terrier so that he can give it a kiss; it doesn't want one. Fred is loudly insisting that he choose the yellow ducklings and, 'Can I please just cuddle one now?' I'm trying to remember if we still have a water trough at home that is small enough for ducklings. A long queue is forming and Charlie, who didn't really want any more animals to look after anyway, has lost the will to live.

Still, nothing is too much trouble for Annie, who owns the place. I think she started it as a hobby and is now run off

I turn around to find that Gus is staggering under the weight of a large brown paper sack of grain.

her feet; she never stops smiling as she catches chickens with an authority and speed that should be recorded for posterity and manages me and my horde admirably. In she clambers, over the gate with one boy at a time, and silence falls as they focus on the serious business of choosing our new pets. Into the cardboard box go two yellow Aylesburys, their massive feet waddling, they surprise Fred with their speed. Gus selects two Khaki Campbells. 'These are fantastic layers,' says Annie and then the sweaty palms of Arthur barely grasp the delicate bodies of two guinea fowl before they are whisked away and added to their new friends. After a few startled moments they immediately spot the Aylesburys at the other side of the box, identify them as potential big fluffy mammas and snuggle up to them.

All are now happily installed in the barn at home, in a deep pile of clean

shavings and under a red heat lamp. They will stay here for the next month until they have grown big enough to taste outdoor life. Fred continues to watch over them, I have taken out his lunch but despite his insistence that he wants to sleep there with them I am going to force him in for dinner.

Back garden ducks

Questions to ask yourself before you take the plunge:

▷ **Do you have time?**
Ducklings need to be checked a few times every day, they need to be kept warm and dry and have food and clean fresh water on hand at all times.

▷ **Do you have somewhere outside to keep them?**
You will often read that ducks don't really need water but what is a duck without water? Ours were in the stream within three minutes of us putting them outside in their new home and on hot

days they often spend all day in there and on its banks. A small stream is fabulous because there are plenty of opportunities for foraging, and the flowing water is always kept clean naturally. However, a small pond, or at the very least a baby bath or paddling pool, is fine and fun. It's not just the ducks which will benefit, we are about to install a bench in our duck enclosure because they are such fun to watch in the water.

Protected by an electric fence the ducks are safe from foxes and otters.

An electric fence and a house for the ducks to go into at night are essential if, like us, your neighbours include foxes, badgers, mink or otters. They need some sunny places and some shade.

▷ **Do you have somewhere inside to keep them?**
For the first weeks of their lives ducks need somewhere warm, sheltered and draught-free to live: like a garage, shed or stable – somewhere you can board off a large corner and hang a heat lamp. They need enough space for a food dispenser and water, and for them to be able to move away from the heat so that they can moderate their temperature. The floor needs to be covered with shavings and newspapers because they will make a watery mess!

As the weather gets warmer they can go outside during the day – we used an old chicken run to keep them in. It is great to watch their first moments on grass and they soon get used to their new-found skills of grazing and sunbathing.

Captive ducks won't start producing their own oil for their feathers until they are three to four weeks old so don't introduce them to water for swimming until they are at least a month old and then a warm day is best.

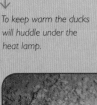
↓
To keep warm the ducks will huddle under the heat lamp.

● REC

Charlie: Some may say the moorhen is not the most exciting bird to photograph but I would disagree. Like all birds, if you spend time watching them you get to see things that are interesting and beautiful.

The moorhens on our river are timid and difficult to get close to – I have never had a huge amount of luck photographing them. One of the problems is that not only are they quite smart but they really know every inch of their territory, so when you dump a great big hide in the middle of it, they will often avoid it for a day or two. I had this problem one year whilst trying to film them at their nest. They simply would not tolerate the hide too close and I had to sneak it nearer to the nest at a painfully slow rate every few days until I could get close-up images, and even then they watched me cautiously the whole time.

Almost every village pond, canal, reservoir, lake and river in southern Britain has moorhens pecking about, and more often than not they are so used to seeing people that they are quite approachable. Places like these are going to be the best sites to concentrate on.

There is no better position for photographing a moorhen than being down low with it.

↓
A great time to photograph moorhens is late spring when they're foraging a lot to feed chicks.

Moorhens are most active between late February and May. In the early spring the females will battle each other for males. These fights can be very aggressive and make great photographs. Often they occur in water with the two birds slapping each other in the face. Sometimes the birds end up on land in the same position and often they will roll around pecking and jump at each other, slapping their opponent's face with their large feet. At the right time of year in the right place these fights can be seen a lot.

My suggestion for photographing them is to get low! There is no better position for taking pictures of a moorhen than being on the deck with it. So if that means lying down, then get some old clothes and lie down. The difference between lying down and standing up is that you're hugely noticeable. If you're using a digital compact, zoom in as much as you can and lie still, allowing the birds to get close to you. You will never get them really close up by chasing them. If you've got a DSLR with a zoom or telephoto lens do the same and set the camera to a fast shutter speed – preferably above 250th of a second. Although the fights are fast the birds don't travel far, especially when they are locked together, so the auto focus should be able to cope fine. If you've got a fast motor-drive on your camera, wait until the action is at its best then 'machine gun' it! The more frames you can get during the action the better, you can delete some of them later.

Photographing moorhens on the nest is also possible, but make sure that you work with birds that are not too wary of humans. Places such as the Wildfowl and Wetlands Trust at Slimbridge are great for taking pictures. The birds are so tame that they take almost no notice of humans. Whatever you do, make sure that the birds come first, and really watch their behaviour. If they are not approaching the nest because you are too close then get back; no shot is worth a bird abandoning its nest. Eggs left exposed without an adult bird on them can die very quickly or get stolen by gulls and crows.

Fledgling moorhens are also great to photograph as they have such a strange pallet of colours on their heads. If you can get shots of the parents feeding them, even better. One thing I have learnt from working for *National Geographic* is that pictures of animals are generally better if the animal is doing something. The shot doesn't even have to be pin-sharp or perfectly exposed, the story it is telling is more important. A shot of an adult moorhen feeding a chick is far more interesting than just a straight portrait.

Moorhens are great birds to photograph – they can be found almost everywhere and are often very approachable. Just do me one favour – get down to their level!

↓
A telephoto lens and fast shutter speed will freeze the action.

Halcyon River Diaries

Late Spring

Charlie: It's 8:45 p.m. and Dusk. I'm squatting under a small bush on a mound overlooking the Manor pond. It is raining hard and I have come out completely unprepared – no rain gear and nothing to protect my camera and lens, so it's all getting soaked.

Hidden in the reeds of a small island that sits in the middle of the pond, I can just make out the dark shape of a female mallard duck. She has bedded down for the night with her ten ducklings; she has enveloped them beneath her wings to keep them warm and safe.

A fox appears from the woods behind the pond and begins to trot towards it. He picks his way through the Jacob's sheep who lie sleeping in the wet grass and reaches the reeds that skirt the edge of the pond. He sniffs the air and listens – he is searching for ducklings! The mother duck is in trouble because she is upwind of the fox and he can smell her. He starts to move round the pond towards the island, nose held high to locate the scent of the duck family. I try to keep track of him through my lens as he comes closer but he spots me. He immediately runs up the hill away from me but then stops suddenly and stares. With some distance between us he relaxes and sits down to watch me.

The rain becomes heavier and I'm beginning to think that sitting out in it is not such a great idea. However, I really want to stick with my mallard and find out how she copes at night in a world where just about every predator is trying to eat her babies.

Darkness finally falls. My infra-red camera works well at dusk as it is very sensitive to light, but in the dark it is useless. I turn on the two powerful hunting lights attached to my filming rig. I have adapted these by replacing the glass on the front of them, with red filters. Now the pond is drenched in red light and I can see again.

A duck on the other side of the pond becomes restless. I focus on it with my camera. It begins to quack frantically. Something is worrying it. I scan around for the fox but can't see it. The duck erupts out of the water quacking loudly; I scan the area with my lights and camera but can't see anything. Other ducks join in, appearing from the reeds. The ducklings vanish and scatter. My heart starts to race – I know something is hunting them, but what?

Within just a few seconds a tranquil night on the pond has descended into

● REC

chaos and then I find out why – an otter! At first all I see is its sleek back as it rolls in the water but a minute later I get a proper view as I spot the otter sneaking through the reeds on the edge of the island, going straight towards the ducklings. I scan the camera ahead to find the ducklings but they are gone. The otter vanishes into the reeds and the only way I can follow it is by watching the tops of the reeds moving – he is one sneaky character! I lose him for a while.

Meanwhile all the ducks and ducklings have gathered into the middle of the pond and they are calling nervously to each other. Where is the otter?

I pick up ripples on the edge of the reeds and a very brief glimpse of light shining from the otter's eyes. The ripples move out towards the ducks. They quack nervously and then all the mother ducks suddenly erupt out of the water quacking. The otter pops up right

The otter is the ducks' worst nightmare and the ducklings didn't stand a chance.

Within just a few seconds a tranquil night on the pond has descended into chaos.

beneath one of them a split second too late. The ducklings scatter in the reeds and their mums keep calling. A moment later they erupt again and once more the otter pops up underneath them. This happens a few more times. It then dawns on me what is going on. The mother ducks are trying to save their young by drawing the attention of the otter to themselves – a very dangerous game. But it seems to be working; the otter is after duck, not duckling!

The chaos on the pond goes on for well over half an hour before silence finally falls. The ducks start to regroup quietly and move back towards the reeds to hide. I scan the pond with my camera but can't find the otter. I film the ducks for a while as they begin to settle, and I try to dry some of the rain from my camera and lens.

Ten minutes later the ducks start quacking again. I spot some ripples and spin my camera round to scan the pond. Two eyes beam back at me like headlights. In the water by the island lies the otter – looking straight at me! I've been spotted. The otter rolls in the water and vanishes.

Philippa: It's just me, the hairy men and a camera crew in a ditch looking for poo, tiny poo.

My old biology teacher, by rude reference to the fact that most biologists have lots of hair and are bearded (well, mainly the men), used to call male biologists the hairy men and it has kind of stuck in my brain. Except these aren't just any old hairy men, they are the men from the wildlife trust who have turned out to be really helpful – they were so dubious about the thought that we might reintroduce the water vole without any experience, that they immediately offered to send people to help with some initial research. It wasn't quite the motivation I was after but I do need help to find out if we have a resident water vole population. They try to do regular surveys (finances permitting) but there have been none on our river for a while, so we agree to go to a few areas where water voles were spotted many years before.

So I am joined by James Field, sporting rucksack, compass and clipboard and lots of field craft; and Robin Marshall Ball, sporting long hair, a beard and a hilarious sense of humour. I like them both immediately and we set off in the car laughing and joking and talking about other conservation projects.

Soon we are in the first of these areas. Strictly speaking it isn't our river but a ditch which runs into it, near the source. We park in a lay-by and cross several fields tracking the ditch as we go. I have started striding out in my typical stupidly optimistic mood, confident that we can make this happen and that there is nothing we can't do to bring water voles back to thrive on this river.

'Ah!' Robin's first call of the day is not what I was hoping for. There is a small squeak as some of the air leaks out of my balloon of optimism. I trudge up the ditch – not an easy process when wellies are sinking fast – to a flat muddy patch on the bank. Robin thrusts a finger at the small pointy tracks which James verifies: 'mink!'

James is quieter, he shakes his head, and carries on up the ditch looking a bit depressed. I can tell he is not hopeful. They must think I am a complete idiot insisting that we come all the way out here.

'This is not good news, a mink can finish them off,' says Robin, 'not good news at all, I'm afraid.'

TV and conservation have become entwined and I have to wonder if I would

Looking for water voles means getting down and dirty.

↓

Water vole runs are not easy to spot without an expert eye.

have gone to the extra effort to find out about the water voles in our river if it weren't for the TV project. I would love to think that I would be here anyway, hoping to make a difference, but I'd probably be working on something else to pay the mortgage.

As I gloomily climb the bank after delivering a dismal piece to camera I reflect that poor James and Robin are probably feeling the same about the TV side of things. All the wildlife trusts and practical conservationists are really up

'This is not good news, a mink can finish them off,' says Robin, 'not good news at all, I'm afraid.'

against it, there is very little money available, and so many important projects are put to one side because as a society we don't always place conservation very high up the priority list.

When there is so much work to be done, I don't want to waste their time but we all know that the extra exposure given to the subject by our programme will be well worth their time, and that water voles have no future without public awareness. That said, they know the need to gently reign in the rampant optimism of a TV presenter who, although she may have studied the subject, has none of the essential practical experience of water vole population management which would be needed to successfully manage a reintroduction programme.

None of these things are said, we all smile and joke.

'More mink tracks,' says Robin.

'The other thing,' says James, 'is that this is no longer great habitat for water voles. When they were doing well here it wouldn't have been so overgrown. They need greenery to eat, open banks that are not too steep, not wooded and dark like these.'

He points at the muddy steep sides of the ditch, there are what seem to be rat runs. 'These are possibly old vole runs,' he says, 'but they obviously haven't been used in a long while, there is no cover here and nothing for them to eat and I'm afraid the mink has probably long since finished them off.'

Jim and Robin move along, mumbling and still looking for signs, but I'm sure they have already made up their minds that there are no water voles here.

I follow behind at a distance, I may be looking at the bank but I don't see what they see. Suddenly I realise what it is that

↓
*A stem cut at a
45-degree angle
means only one thing!*

is making me so fed up. I so want to make a difference to this river that I am privileged to live beside, to do something positive and to film it so that other people might be inspired to do the same in their local river. But I realise how naïve I am. People like Robin and James work in conservation every day and the truth of the matter is that you can't just swan in and make a big difference. It is a long hard slog. It's about chipping away at the coalface so that you can see a genuine difference in ten years' time. And in the meantime it's about having to manage all the other stuff, like over-enthusiastic TV producers and local landowners, public perception and budgets. Conservation in the real world is about consistency and longevity, really knowing your subject and accepting that change can't occur overnight.

There isn't anything wrong with what

I want, it's just not realistic.

We round a corner and find ourselves facing a small timber bridge. So, grunting and groaning and leaving wellies behind in the mud we pull each other out of the ditch. I quite expect that this is the moment when one of them will say: 'Well that's it, well done, but nothing doing. Let's go for lunch.'

Instead, Robin says 'Oh this is more like it.'

'Oh yes,' says James.

I follow their gaze down the field where the ditch meanders, devoid of vegetation except for one tree.

'Oh this is much more likely,' says Robin and dives in (not literally). We see nothing but the top of his head, and James takes a few moments to explain.

'You see, here the banks are sloping at the right angle, they are clear with short vegetation and plenty of greenery, this looks much better.'

I can feel myself getting excited again and I dive in after them.

'James, look at this!' comes the call. Robin triumphantly holds up a green stick about an inch long.

'Look at the end of that,' he is bursting with pride. Sure enough it is cut at a perfect 45-degree angle as though someone has sliced it with scissors, a classic sign that a vole has been eating it.

'And it's fresh,' I add. I can't keep the glee from my voice.

James, being more scientific, admits this is good news but insists that we can only really be sure if we find fresh poo.

'Water voles have latrines, little scrapes in the side of the ditches where they go and we need to find one.'

That is enough for me, like a spaniel let off the lead I am off up the ditch, head down, bum in the air. And this media tart, this amateur conservationist, who needs to pick her kids up from school in

↓
Water vole latrines work as markers to other water voles, giving them information about who's on the block.

forty minutes is back dreaming about becoming Gerald Durrell again.

Then, eureka, water vole poo!

The fireworks explode, the crowd goes wild, everyone throws their hats in the air, James even smiles.

And the more we look the more there is: latrines, burrows, tunnels, loads of leftover food, grazed bits. There is evidence of water vole life everywhere. Well, just in this little hundred yard length of ditch anyway.

We are all delighted, this is so significant. It means that to reintroduce water voles here is wrong – now we must focus on protecting the population that is already here. It means that we have to make a plan; we need to find out who owns this land, we need to really worry about that mink. It means that a species that belongs here still exists and that we have found water voles before it is too late. Not overnight, but possibly little by little, by chipping away at that coalface over the next ten years, we might even get their population right back up to where it should be.

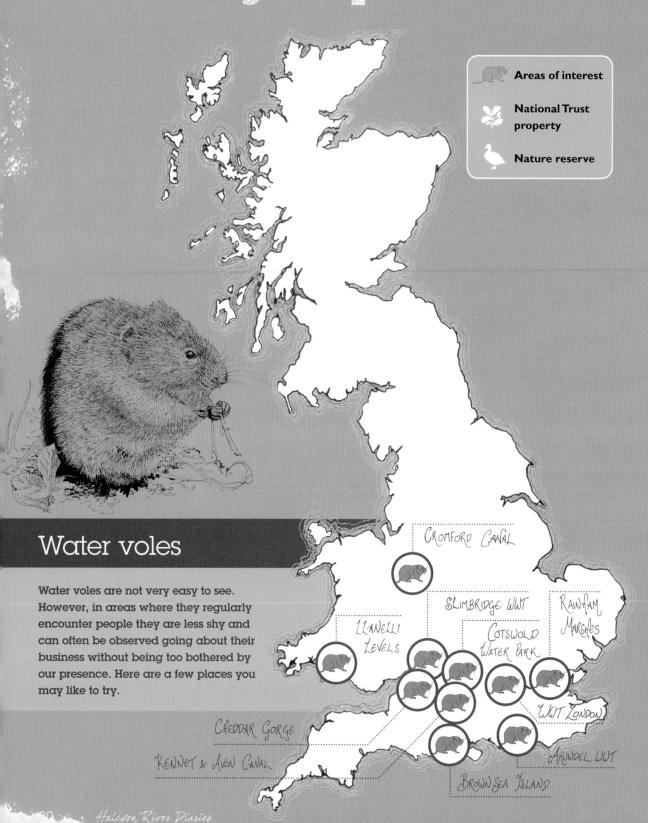

Water voles

Water voles are not very easy to see. However, in areas where they regularly encounter people they are less shy and can often be observed going about their business without being too bothered by our presence. Here are a few places you may like to try.

Legend:
- Areas of interest
- National Trust property
- Nature reserve

Map locations:
- CROMFORD CANAL
- SLIMBRIDGE WWT
- RAINHAM MARSHES
- LLANELLI LEVELS
- COTSWOLD WATER PARK
- WWT LONDON
- CHEDDAR GORGE
- KENNET & AVON CANAL
- BROWNSEA ISLAND
- ARUNDEL WWT

CROMFORD CANAL, DERBYSHIRE

Cromford Canal has to be one of the best places to see these beautiful mammals. The landscape is stunning – like a scene from *The Wind in the Willows*. The voles breed and live among the busy paths and walkways – they are used to people so are fairly approachable.

SLIMBRIDGE WWT, GLOUCESTERSHIRE

Slimbridge has a boat safari, which is great if you want to photograph water voles as you are on the same level as them. The backdrop of reeds and water works well. If you're lucky you can spot the water voles from the car park before even entering the main reserve.

LLANELLI LEVELS, WALES

This area of low-lying land is dissected by numerous ditches and as a result is packed with water voles. In the middle is WWT Llanelli which also has a very healthy population of water voles and as the site is contained and free from mink the water voles can sleep safely at night!

COTSWOLD WATER PARK, GLOUCESTERSHIRE

The Cotswold Water Park covers forty square miles of Gloucestershire and Wiltshire and is home to a wide range of wildlife, including water voles. Water voles have been expanding here since 2002 and can now be seen on many rivers and streams in the area. One of the best places to see them is the River Thames. Walk along the Thames Footpath from Somerford Keynes to Cricklade and you should be rewarded with a sighting.

RAINHAM MARSHES, ESSEX

This is a key site for water voles in Britain and boasts a very healthy population. The RSPB took control of the marshes a few years ago and have enhanced the habitat by creating a number of water courses. Recent population surveys of the voles in the marshes suggest that numbers have increased. So now is a great time to visit.

WWT LONDON

The Wildfowl and Wetlands Trust (WWT) have a centre in Barnes which is a great place to see water voles. Two hundred and fifty of them were released in 2001 and the numbers have increased dramatically since then (between three and four thousand are estimated to live there now). So if you're in the capital it doesn't get much better than that.

CHEDDAR GORGE, SOMERSET

Not far from the town of Cheddar lies the famous Cheddar Gorge, a stunning slice of geology that seems completely out of place in Somerset. The River Yeo flows out of the gorge and has a good population of water voles. Due to the large number of tourists in the area the voles have become very tolerant of people, making Cheddar a great spot to watch them.

KENNET & AVON CANAL

The Kennet and Avon Canal hit the news in 2009 after a survey by British Waterways found a large rise in sightings of water voles on the canal which stretches all the way from Bath to Reading. I have visited the Bath section and it is beautiful – although I've still failed to spot a water vole. Good luck!

ARUNDEL WWT, WEST SUSSEX

WWT have a special interest in water voles and populations of them are being nurtured at all of their reserves. Arundel is no exception and since their reintroduction in 2005 the water voles have spread throughout the reserve. The best way to see them is on the boat safari.

BROWNSEA ISLAND, DORSET

Brownsea Island in Poole harbour is a great place to visit. Not only does it have a very healthy water vole population but they can be easily seen from its reed-bed boardwalk. Brownsea also has populations of red squirrels and sandwich terns – so if you get bored of the voles you can go looking for these instead!

Charlie: Otters are just about the hardest creatures on the river to photograph. On our river they are strictly nocturnal and very secretive which makes photographing them even more of a challenge.

My mate Allan – one of the best otter spotters around.

In certain parts of the country, such as Shetland and some of the western isles of Scotland, otters are more approachable and often stick to a daytime schedule. These otters still pose their own sets of problems, though, and getting close enough to them to get really good shots takes a bit of time and practice.

I am obsessed with otters and have been since I was a boy. To me they are the ultimate challenge for a photographer – they are shy, cunning and clever. They have an amazing sense of smell and great hearing. But they do have atrocious eyesight which is the chink in their armour. This means that if you're quiet and downwind of them they won't know you're there, allowing you to get great views and photographs. This is most true of the otters in the Scottish islands. You can get great views of these otters by simply sitting quietly on the shore and waiting for them to swim past. This is not so true of the otters on southern rivers. These guys pose a much more complex set of problems to the photographer, but problems that ultimately make them a far more attractive proposition. The two methods of photographing otters could not be more different, so I am going to discuss them both.

Photographing otters during the day

There are certain places in Britain where otters are regularly seen during the day. Shapwick in the Somerset Levels, for instance, is currently a good place to see them at all times of the day. These otters have a degree of tolerance of people which makes life easier but they may also be a good distance away and

Otters are just about the hardest creatures on the river to photograph.

getting good shots of them requires a powerful telephoto lens. Wherever you are photographing otters during the day, the rules remain the same. Firstly decide on your location. You are unlikely to just walk out with a camera and get great otter shots. You will need to research a general location and stick to it. Once you have decided on your location, you have two options. You can either walk the stretch of river or coast looking for the otters or you can find somewhere comfy to sit and wait. Personally I do both. You have to be very conscious of the wind direction, though. If you are upwind of the otter you are not likely to ever see it. It will smell you and hide from you long before you ever reach it. When I'm working a coastline or river I always take binoculars. I generally walk very slowly and scan constantly. In rivers I'm looking for ripples, if I see them I try to confirm their source. Most of the time fish and waterfowl are responsible but just occasionally an otter will pop into view.

Once you've spotted your otter you need to get into position. If the otter is in a river it will generally be swimming either directly up or down it. The trick is to get ahead of it, find a spot and settle in it before the otter reaches it. This can involve some fast running. Otters move surprisingly quickly so you need to get a good 200 metres in front of the animal. This will give you time to choose your spot, catch your breath, set up your kit and find the otter.

The longer the lens you can get on the front of your camera the better. It is much easier working with an otter 30 metres away than one 5 metres away, so try to keep as much distance between yourself and the otter as possible. As the otter approaches you need to get your noise and movement down to a minimum. Otters may have bad eyesight but they can detect movement and will react to it if they don't know what it is. Otters will also react to the sound of a camera shutter. Some will take fright and vanish, others will snort in alarm and some will swim over to investigate the noise. The best otter is the one that ignores the noise – an otter behaving naturally is going to be the one to photograph. As the otter swims past, really watch how and whether it is reacting to you. If it looks nervous stop snapping, you don't want to scare the animal in order to get a shot; if it is relaxed keep snapping. If you're photographing otters on the coast the method is much the same, the otters may go out to sea more but they will often come back ashore and generally be heading in one direction.

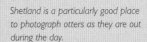

↓
Shetland is a particularly good place to photograph otters as they are out during the day.

Photographing otters at night

Photographing otters at night is a different ball-game altogether and one that relies on camera traps. Camera traps come in all shapes and sizes but they mostly do the same thing, they fire a beam of infra-red light between a transmitter and a receiver which triggers a camera when broken. They are not too expensive and can be attached to any camera with a cable release socket. I use Trail Master camera traps. These are made in the USA and are generally regarded as one of the better systems, not only are they very accurate but they last for months on one set of batteries. Camera trapping is quite addictive, mainly because you never quite know what you're going to get – you could be out camera trapping otters and get cracking shots of foxes or deer, for example. It is, however, fraught with problems and failure so you have to be very persistent if you want to get good results. I am currently doing a project with otters; my success rate is very low. I put two camera traps out every night and I get on average one shot of an otter every ten days – 50 per cent of the time the otter is looking in the wrong direction or the shot is out of focus!

Firstly you've got to know your otter field signs (see 'How to track an otter' on page 34). What you're after is either a run regularly used by otters, or a spraint point. When you've found your spot you need to consider a few things: which direction the animal is going to be travelling in when the trigger fires, where you are going to put your camera, where you are going to put your flashguns and is all your kit going to get stolen?

Otters will very often travel in one way up a river and another way down it. They may spraint on a specific rock when they're going upriver but ignore it on the way down, for instance. It is quite difficult to work out which way they are going but useful signs are footprints. It is important to try and work this out so that you can place your camera in the right spot – you don't want to photograph the back of an otter's head! When you've decided where to shoot from you need to get your camera in. Otters really know their territory, if you whack something new into it they will notice it. So you need to get your camera low and hidden. I hide mine in grass under a waterproof cover. Photographing otters at night means using flashguns. This is the only way you are going to be able to light them. One flashgun will suffice but I tend to use two. If you don't have a flashgun you could use the camera's onboard flash. This will work but not give you the best lighting. The same is true for putting a flashgun on the hot shoe of your camera. Light fired directly from the camera is not always preferable as it is usually fairly flat and harsh. The best system is a flashgun attached to the camera by a cable or wireless unit that is either off to one side of the subject or above it. I almost always use the same system when camera trapping. I have a flashgun fairly near the camera on a wire that fires a fairly small amount of light and I have a second flash mounted above where I hope the otter will be which delivers most of the lighting. This system gives a nice atmosphere to the shot.

Once you have placed your camera and worked out your lighting you need to set up your infra-red beam system. You don't really want your beams in shot so move them out of the frame. Some are powerful and can be moved a good distance from the subject. The key is to make sure that they fire the camera when they are broken at exactly the right spot and that spot has to be in focus. So make sure you have your framing and lighting and focus all set up first – and make sure your camera is set to manual focus! Otherwise you will probably end up with a blur.

Once you've done all these things you're good to go. I generally leave everything powered up all night. The camera battery and flashguns can just about cope with a night of it before dying. Leaving everything on means that the instant the beam is broken the shot is taken. Camera trapping is annoying and fiddly but ultimately great fun. Good luck!

↓

Most camera trap systems rely on an infra-red beam being broken – the unit that puts out the beam then fires the camera.

I put two camera traps out every night and I get on average one shot of an otter every ten days.

Philippa: I am a Hampshire girl born and bred, I grew up playing in chalk streams. One of my earliest and happiest memories is of being a toddler, perhaps my first halcyon day. I remember it keenly, a hot summer's day, a picnic, the safety of my mum close by. But I remember it mostly for my encounter with the river: cold, clear, shallows but with darker bits which even at that age I knew might get deep. Was that instinct? I can clearly recall liking the gravelly bottom, the thrill of the cold water on my skin. The overwhelming sense that this river was a living thing, not like any other water I had played in.

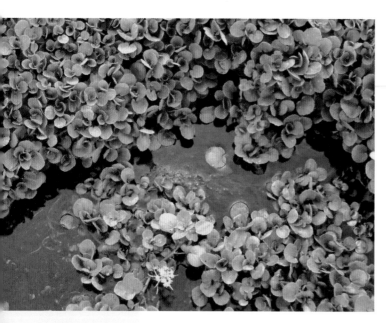

↓
Watercress thrives in the pure chalk streams of Hampshire.

As I grew, bike rides took me further afield but I always stopped to rest beside a lazy stream or cross over small bridges with babbling brooks, waiting long enough to spot the dark torpedo of a trout moving just enough to maintain its position. Then, grinning with the satisfaction that I had spotted him hiding, I would push off on my bike.

Part of the surroundings that I cycled by and took for granted were the watercress beds. If you grow up near Winchester as I did you can't really avoid them, the watercress farms are simply part of the landscape. I had no idea that they were so special then.

Hampshire's chalk downland provides the perfect geological conditions for growing watercress. Over many years rainwater percolates through the chalk, absorbing minerals, and then babbles out from deep underground springs, to pour into a series of man-made runs, some of which have been in operation for over a hundred years. The water is constantly flowing, directed over huge gravel beds where the young watercress is grown. In a matter of just a few months it goes from seed to shop. First it is thrown onto the gravel beds when it is the size of windowsill cress. It willingly takes root and grows quickly.

On the day we film at the watercress farm the heavens open as if to obligingly demonstrate the never-ending water cycle on which the watercress beds so rely. I interview the farmer, wincing against rain so forceful that it penetrates my mac and drips off the end of our noses. We bend down and pluck a fistful of deep-green watercress from the cold moving water, the taste on this drab grey day brightens my whole mouth. It is of course vaguely familiar from packets bought in the supermarket but they don't bring my senses alive like this, fragrant and peppery and with another layer that seemed to come from the water itself, this is watercress so fresh that it is like a different product.

Being a grow-your-own type, I am immediately inspired. Surely for growing a waterside plant that naturally inhabits streams and rivers I live in the perfect place.

Well, it turns out that I don't. As it absorbs the good minerals from the spring water, watercress will happily and readily absorb any pollutant so whether there are cows in the river upstream or chemicals running into the river from a farm, the watercress will soak it up.

Here the purity of the watercress is guaranteed, springwater comes straight from a borehole and is regularly tested.

And before you are inspired to collect your own from the wild beware of the liver fluke (*Fasciola hepatica*). These are cysts on the underside of the watercress leaves, when eaten they then burrow through your intestines and into your liver and bile duct where they form adult flukes as long as three centimetres.

So this is possibly the best place to be picking and eating watercress, and once I have started I find I can't stop. Neither can our crew – for days we have it in sandwiches and soups and really enjoy the fresh taste. And so we should, watercress is a superfood, cancer fighting, rich in minerals.

The nutritional value of watercress

Watercress is a low-calorie food but particularly rich in vitamins A and C and with folate, calcium, iron and vitamin E.
Its potential in fighting cancer is particularly exciting with studies in laboratories indicating that watercress might reduce prostate and breast tumours and perhaps has beneficial effects against smoking-related cancers and against colon cancer.

Watercress and feta omelette

This makes a lovely quick garden lunch for one when you have half an hour to yourself to sit and enjoy the sunshine. It takes minutes to throw together but because you have cooked something for yourself it feels like an indulgence.

1 handful of watercress, finely chopped | 50g feta cheese | 2 eggs | butter

You can use a frying pan but there is nothing quite like a little omelette pan. Mine is a heavy old red one, it feels good to use which starts the joy of the meal just a little earlier.

While the butter is melting in the frying pan, beat the eggs. Just as the bubbles have disappeared on the butter, pour the eggs into the pan. When the omelette is almost set crumble the feta on it using your fingers.

When the feta is just melting add the watercress in a heap in the centre of the omelette and without further ado fold it and serve.

The crunch and bite of the watercress remains against the gooey saltiness of the feta and blandness of the egg.

Sit in the sunshine next to your favourite plant and enjoy.

Watercress should be celebrated and in Hampshire we do that in style. In Alresford the people turn the beautiful old town over to celebrating watercress once a year in May. It feels like going back in time, like stepping into a costume drama. We walk up the main street behind the Watercress King and Queen, children bedecked with watercress sitting high on top of a horse and cart, they toss the first of the harvest to us serfs. Morris dancers wave their hankies, a brass band plays and the farmers' market is a feast for the eyes. As we are swept along by the crowd I can't wait to revisit the stalls, their legs groaning with the weight of local produce, as soon as the parade is over. Our children clap and sing and even Charlie gets caught up in the spirit of it.

So much so that I have to wait a little longer than I had hoped, because Charlie enters the world watercress-eating championships. This turns out to be riotous, with all the children in the audience screaming for their fathers who are all standing in a row, dignity lost, cramming their faces with green leaves and chewing desperately. I'm unsure whether to be relieved or sad that Charlie didn't make it to become the world watercress-eating champion, but I haven't laughed so much in ages.

It was a day to forget everything else and simply celebrate a local watery product, bearing in mind that a bunch of watercress is also reputed to be a great cure for a hangover.

Interesting website

For more information on this versatile and nutritious plant go to
www.watercress.co.uk

Comforting watercress soup
with squash and sweet potato

Although it doesn't immediately spring to mind, the spiciness of watercress is a great complement to the soft comfort of butternut squash. They make a great partnership for all sorts of dishes, from pasta to risotto, but this soup is a particularly lovely one.

**I large butternut squash | I medium sweet potato | olive oil | I bag watercress
I litre vegetable or chicken stock**

Roasting the squash and potato may seem a bit of hassle but it actually saves a lot of peeling. It also gives the soup a richer flavour.

Chop the butternut squash into quarters, remove the seeds but leave the skin on. Chop the sweet potato into large chunks, again leaving the skin on. Liberally cover with olive oil and roast in a hot oven until soft (200°C/400°F/Gas 6).

When they have started to brown remove the sweet potato and squash from the oven and leave to cool for a few minutes. Pick the quarters up and use a knife to slide the soft orange flesh of the butternut squash from its skin.

Place both in the blender with roughly a litre of vegetable or chicken stock and blitz.

Heat up in the saucepan and when the soup is almost ready stir in the watercress.

Serve with bread or oatcakes.

… but the very best thing
of all to do with watercress
is to sandwich it.

Philippa: It seems impossible to believe that it is already May. The hawthorn is in full flower like some architectural feature in the garden, a frozen fountain of flower sprays. Spring was so full this year and now when I step outside I can feel the garden groaning with growth. The novelty of the kind sunshine makes us look at the detail of the season. The beech tree's tentative leaves are like babies' gloves glowing with the new softness of their green, yet they make me think ahead to autumn when the same leaves will grow orange as the sun rises behind them and the mists rise from the falls below, such is the deep joy of getting to know a place so well.

The focus has moved from spring flowers in the borders to the meadow buttercups on what should be lawn but which I refuse to cut. We are mowing pathways through these wild flowers and leaving huge swathes of lush green which is becoming more and more interspersed with yellow and, when you take a second look, the pale pinky-white flowers of I don't know what.

I have been trying to cater for the needs of the wild flowers for two years now and it is paying off. I can spot cowslips where there were none and others which I don't know the name of – I need to get down on my hands and knees and identify some of

them. It has dawned on me that flood meadows are rare and our children are playing on one every day. I just need to let it be a flood meadow instead of a lawn.

Where they would be playing football the children now run down the paths between yellow flowers. As I wander and wonder with the camera, Gus follows my lead and he follows behind me, taking photos.

As a child, one of the nicest memories of growing up in Winchester was of long walks along the water meadows, sometimes a little soggy underfoot but always with something pretty to look at. The most sublime being the wonderful vision of yellow that arises at this time of year, best on days when the sky is

blue. Now, as an adult right here in my back garden, I know it is a fragile habitat but I have very little knowledge about the flowers I can see today. Inspired, I determine to find out more.

↓
Taking photos of wildflowers is a great way to discover the beautiful detail of them.

Charlie: When I turned up this morning I was expecting to film kingfishers' courtship feeding outside their new nest. What I've actually spent the day filming is some of the most hectic and amazing kingfisher behaviour I have ever witnessed – the problem is I have absolutely no idea what is going on!

↓
Kingfishers protect their territory aggressively.

I do know this – there are a male and a female kingfisher trying to dig a nest together in the mud bank opposite my hide. However, there is another kingfisher hanging around and causing the pair big problems. This other bird could be either another male or another female. At this time of the year there is a flurry of newly-fledged kingfishers on the river – these young birds often cause problems to adults as they try to find themselves a nice stretch of river to live on. Kingfishers are highly territorial and won't put up with others trying to move into the mile or so of river that they hold as their territory. At first I thought this was what I might be witnessing – a young bird looking for somewhere to live. But I soon realised that this wasn't the case; this bird behaved like an adult – fast, accurate flying, aggressive, confident and with the distinctive high-pitched adult whistle.

Could there actually be two other birds hassling the pair? This was a strong possibility. This nest I am watching has a long history of aggression surrounding it. Good nest banks are rare and there just happen to be two very good ones 150 metres apart here. Generally a pair of kingfishers will nest in each bank and their proximity to one another always causes endless fights as they antagonise each other.

I settle on this theory for a while – I like to know what's going on, I feel uneasy when I don't. The nest bank that the resident pair are digging in is about six feet high, above a small rapid. Outside the nest hole there is a nice curved branch that hangs down from the root system in the ground above the nest – this is where the kingfishers sit when they are digging and courting. I have my camera focused on this perch as it is the centre of all the kingfishers' attention. The male bird is sitting on the perch whilst the female is working inside. A series of loud aggressive whistles comes hurtling down the river and the turquoise streak of a kingfisher speeds into view. The male bird on the perch stands and turns to face the other kingfisher who flies right at him. He whistles angrily as he is knocked off. Like lightning he rallies himself and is in hot pursuit of his aggressor. I try to follow the two with my camera as they speed off down the river but it's useless – they move too fast for me.

Thirty seconds later they're back, firing straight up the river, whistling furiously. They zip over the top of my hide, clipping

If neither bird will yield,
killing the opponent is
the only option.

it as they go, and they're off upriver again. I grin to myself – I love a fight! A few seconds later the male swoops in from behind the hide and lands back on his perch. He whistles and the female pops out of the nest. The two sit together whistling to each other – half aggressively, half to confirm their bond. Kingfishers never really enjoy each other's company, even when they are a pair.

The relative harmony doesn't last long, though, the aggressor is back. This time silently dropping out of the trees and knocking the female off her perch. In a flash and an eruption of angry whistling all three birds hare off in hot pursuit of each other – who is this bird?

I watch the same behaviour all day. By the afternoon I am convinced that there is only one aggressive bird and the pair involved in this dispute. I stick with this new theory and then start to wonder who is this bird and why it is being so aggressive.

There is another favourite perch by a small pool surrounded by reeds about thirty metres behind my hide. Here fish gather in the relative calm of the pool. The male is fishing it. He catches a fish and eats it. He then catches another fish, kills it by bashing its head on the perch, then turns it in his beak so that its head is facing forward and then he flies off downriver whistling – this bird has chicks. I already know this, he's been doing it all morning and I've been following his nest throughout the year. This behaviour confirms to me, though, that this is the resident male. The female pops out of the nest after digging and flies upriver to the perch. She immediately catches a fish and eats it. She then catches another and eats it. She's also been doing this a lot today, but she has not once flown off downriver to feed the chicks. The male returns after a couple of minutes and the two sit next to each other and fish for a while.

Out of nowhere another kingfisher appears and lands on the perch next to the pair. It's a female! The male bird doesn't know what to do. The first female looks furious but doesn't react. The intruding female catches a fish and flies off whistling, the resident female goes after her. Now I am really confused.

The aggression is really starting to heat up now. The aerial dogfights seem to be relentless; up and down the river at astonishing speed – I give up trying to film them. Occasionally they stop on nearby perches and square up to each other but I've completely lost track of who's who. My aim now is to film a fight. Kingfishers will sometimes kill by trying to drown each other, this is rare and only happens when arguments between birds fail to get resolved through other methods. I can feel a fight coming on, though, and there are few things in the world of kingfishers more amazing to film than a fight. The problem is that it could happen anywhere – where do I point my camera?

The kingfishers answer that question for me a few minutes later – at the back of the hide, not the front! Two birds screech round the corner and land in some roots overhanging the river. I watch them out of the back flap of my hide. They square up for about a minute before one launches at the other. The two lock beaks and tumble through the roots lodged together. They fall into the water and hop and flap about, neither yielding to the other. I panic – what do I do? I grab the little camera that I use to film myself while I'm filming with the big camera. I power it up and stick it out the back of the hide – I'm panicking and press all the wrong buttons! By the time I find focus the birds part and fly off in different directions. I curse. I hate to miss such great action.

The problem with my hide is that it is small and it is almost impossible to change your filming window once you're installed with chair and tripod. Still cursing, I sit and wait for the birds to return – do I continue looking forwards at the nest or do I reposition to the branch they just fought on and look out of the back of the hide? I opt to sit tight. Big mistake. Within five minutes they're back and fighting again in exactly the same spot! This time I go for it, I spin the big camera round, fall on the floor and half throw the hide in the air. I scrabble to find my feet, hit Record and fumble with the hide and lens – by the time I get a view the birds have split up again!

I spend a further five days sitting in my hide. The kingfishers have incredible stamina and seem to just keep fighting. I never get my fight but what I do work out is exactly what is going on with the birds. The male bird has run off with another woman! Not only that but he is attempting to drive his ex-wife out of the territory, even though they both have chicks together in a nest downriver. To his credit he does continue to feed them alongside his ex, but he is very aggressive every time he sees her. This is the first time I have seen an intruding female successfully break up a pair. I have seen females attempt it before but never succeed. My fear now is for the chicks downriver in the nest. With all three kingfishers at very heightened levels of stress and aggression the poor chicks that are due to fledge any day will be fledging into a warzone.

Charlie: Believe it or not, kingfishers have no blue pigment in their wings. The blue we see is actually light bouncing off them after being manipulated by structures within their feathers.

Kingfisher feathers are designed in such a way that they bounce only blue light back, allowing the feathers to absorb the other wavelengths of light, such as red and yellow. This effect not only explains why kingfishers are never actually one single shade of blue, rather they vary from dark blue to bright green depending on the angle of the light hitting them. This scattering of light by particles is known as the 'Tyndall effect' as it was first described by scientist John Tyndall in the nineteenth century. It is fairly common in nature and if you want to see the very best example of it, simply tilt your head skywards and ask yourself – why is the sky blue? The answer is the same.

Kingfishers aren't blue – they're dull grey-brown!

Halcyon River Diaries

Summer

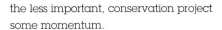
Philippa: We have great success with the water voles this week, we hear from Natural England that our licence to work with them has been approved so it is with much relief and a sense of urgency that I grab Charlie and whisk him straight up to the top of the river to show him the ditch where we had found our resident surviving population.

I'm on a mission. When it comes to conservation one of the biggest obstacles can be other people – they are the key. If you can get people inspired enough to help, and to understand why you need to bother conserving an animal, then most things can be achieved. One of the best ways of doing that is with images. With an animal like a water vole, which is now so rarely seen in the wild, this is particularly the case. Reading about, or hearing someone speak about, water voles is one thing but seeing the creature going about its business in the wild is another.

Getting some good footage of our water voles is one of the most important things we can do to give this small, but none

the less important, conservation project some momentum.

As we park the car and gather up the kit I begin to feel slightly nervous. It has been a few weeks since we came here with James and Robin and I know now just how aggressive mink can be. My anxiety increases as we cross the small stile and walk through the fields. The heavy metal tripod hurts my shoulder, it is so awkward, I shift it but there is never a comfortable position. Just because we found water voles last time, there is no guarantee that we will find them now.

We are quiet as we walk, Charlie carrying the heavy camera, his head full of the other things he needs to do. I watch the pollen waft across the field like mist caught on a breeze. My nose begins to prickle in response to the clouds of pollen, I rub it and, yes, here is the first sneeze, my hayfever has started. Here we go, now there will be nothing subtle about our approach.

We clamber over the stile onto the little wooden bridge that spans what cannot yet be called river, only ditch, and I point along it and show Charlie where we are heading. I'm worried. Water voles are such a tenuous population that it might only take just one mink prepared to go that extra hundred yards to finish them off completely.

We walk alongside the ditch until finally I see the smooth light-coloured rock I left as a marker and drop the tripod with relief.

'Here we are, this is the marker I left.' Charlie looks down into the ditch,

frowning and wincing, I can already sense his scepticism.

I drop down into the ditch getting stung by nettles in the process and bend upside down, holding my hair to make sure it doesn't dangle into the mud. I peer into the side of the bank close to the water, sure enough the latrine is still there and, even better, there is fresh poo and clear footprints where the voles have been using it.

'What are you doing?' says Charlie to my backside.

I stand up to respond, 'Looking for the latrine, it's still there, I'll show you in a sec.'

I reach up and gently part the long vegetation on the opposite side of the bank, revealing more holes with tunnels that stretch back into the field and also lots of runs like open tunnels running

along the side of the bank which weren't there before. I show Charlie, vole 'leftovers', the pieces of vegetation that Robin had shown me last time. Now there are more of them. In addition I am

Filming water voles on small ditches requires lots of kit.

Remote cameras are the only way to film voles in this ditch.

thrilled to see little lawns on the top of the banks where the voles had been grazing.

I look up to see Charlie's reaction. I can tell he is underwhelmed, but not by the voles.

'You really want me to film in that? I can hardly even fit in the ditch, let alone get a camera in there.'

I sneeze in response, raise my hands for help and Charlie pulls me back onto the bank with a big squelch as the mud releases my wellies. I try and resist rubbing my eyes.

'There is no way, even if I can sit there no vole is going to come down that ditch', he says emphatically.

I want to say something helpful but sneeze again, I can't resist it any longer and start rubbing my eyes. Charlie, quite relieved by my unusual silence, uses the time to think.

'The only way around it is to fit some remote cameras next to the latrine, perhaps with infra-red triggers.'

I replace my sunglasses and sniff.

'They should be hitting breeding season at which point the population will spread and so the signs should expand a little up and down the ditch', he continues.

I find an old tissue in my fleece pocket, it has been through the wash and is a bit hard but I am grateful for it. I blow my nose and sneeze again.

'Perhaps if I fiddle with one of the cameras in the workshop, I could ... I'll trail some cables over the field here and sit well away from the ditch ...'

I sneeze three times in a row, all I really want to do is gouge my eyes out but I feel positive, if anyone can get the shot, Charlie can and without me nagging he has already started to work it out.

We return home, him pessimistic and worried about how long it is going to take to get this shot, me optimistically sniffing, full of hope that getting this picture will give us something tangible to show people and perhaps inspire them that here in this quite unspectacular small muddy ditch there really is something worth conserving.

Philippa: We have made an incredible discovery! There is one, just one, Bee Orchid in the flood meadow. It is odd because it is situated right next to the see-saw, of all places. It is beside one of the paths that we mowed earlier in the year.

Any orchid means that the grassland is healthy; orchids rely on a complex relationship with fungus on their roots and for the fungus to be strong there can be no herbicides or fertilisers. So this is great news for the health of the flood meadow.

A Bee Orchid is special, they are not the rarest type of orchid by any means but are really interesting. They have evolved to seduce bees. The flower mimics a female bee in size, shape and colour, inviting a male to mate. And the plant even gives off pheromones to further convince the poor unsuspecting male. Instead of a good time the bee simply comes away with a sac of pollen to pass on to the next Bee Orchid.

Of course this is something we must film, especially since our new beehives are only a hundred yards away. However after scurrying inside and researching I find that we simply don't have the right kind of bee. In fact it doesn't appear that we have the right kind of bee in this country at all.

I wander back out with a cup of tea and stare at my lonely Bee Orchid, never destined to seduce a bee. Always waiting.

I put a metal frame around it so that it can't be knocked over by a dog or football. I inspect the flower again and take some photographs. It is beautifully detailed, exceptionally designed. It doesn't matter how many times I search around, I simply can't find another.

It seems such a shame – all that evolution, all that effort to perfect just the right look and smell that will drive a bee wild, goes completely unnoticed. I sit down in the grass and flowers beside it,

enjoying the evening chatter of the birds. Living as I currently do, a lone woman in a house full of boys and a blokeish-jesting film crew, I understand.

Charlie: I'm sitting in my hide outside the nest of the house male and the intruding female. I've been working this spot for a while, filming the relentless arguing.

↓
Adult kingfishers can be very aggressive towards chicks – even very young ones.

The action around the nest this morning is as it has been for the last week or so – lots of whistling and aggressive flying, with the occasional gap for feeding and nest digging. Every morning when I turn up I expect to see newly-fledged chicks from the nest that the house pair had started together downriver when things were a little less fraught between them. By my calculations the chicks should have fledged a day or two ago.

Several more hours of arguing and nest digging have passed and I'm watching a newly-fledged kingfisher! It is sitting in the sticks that have piled up around the roots of a hazel tree on the bend below the nest bank. It is bobbing in characteristic style. Kingfisher chicks bob continuously in the few weeks after fledging. I have no idea why. The bird is dull in comparison to the adults, the blue is much muted and the

orange chest is much less vivid. The feet are black instead of pink and it has a white tip at the end of its beak. There is also a dull grey-blue band around its chest. Looking at the fledgling well camouflaged amongst the sticks and foliage, I realise perhaps why it is so muted – unlike its parents it doesn't really want to be seen. I get some shots of it sitting quietly on its perch.

The intruding female appears and lands on the perch outside the nest. It is a few seconds before she spots the fledgling and when she does she immediately adopts a threat posture. The chick appears to have no idea what this means and ignores her. The female remains in this position for twenty minutes, silently threat displaying to the chick. It is not until the male appears that things really get going. The male lands on his perch opposite the nest bank and almost immediately the chick locks on to him. At first the chick calls to him repeatedly by making a clicking sound, then it flies over and tries to land on the perch next to the male. The male bird becomes very irritated by the chick and turns on it, pecking at it. The chick tries to remain on the perch but gives up and flies back into the sticks beneath the willow. The female follows with her head remaining very firmly in the threat position. The chick is now much closer to her, about six feet. The female flies and lands on a perch right above the chick, readopting her threat position. The chick clicks at her, assuming her to be friendly; she isn't and launches at the chick, knocking it off its perch and pecking at it as it falls back into the sticks. She then flies back up to her perch and displays, as the confused chick re-perches itself. As the chick finds its feet the female is back on the perch and again the chick is knocked over and pecked.

Kingfishers are very aggressive to intruders and the female clearly thinks that this chick is one and needs dealing with. This worries me – she could easily kill the chick if she chose to and the father doesn't seem to be remotely interested. The chick flies downriver a few yards, obviously shaken by the encounter and settles on a hazel branch. The female remains on full stare. A few minutes later the male turns up with a fish and lands on his perch. The chick springs into life again when it spots him and flies over

to land next to him. The male shuffles away but the chick follows, beak open, begging for the fish. The male flies off downriver and the chick goes too.

From my hide I can see a hundred yards or so downriver and I am getting a lot more sightings of kingfishers there than I am next to my hide. I decide to up sticks and relocate the next morning. When kingfisher chicks fledge they generally find an area and hang around it together. My guess is that the pool downriver from the hide is that spot.

I'm in my hide early the next morning. It is in the water because the banks are steep on the bend. I'm focused on a good-looking perch and hoping that the kingfisher chicks will like it. It's not long

↓

Chicks will beg and bother adults who will sometimes peck them repeatedly.

When kingfisher chicks fledge they generally find an area and hang around it together.

until I find out that they do. I hear the clicks of a kingfisher chick coming upriver and I turn to see three – two chicks pursuing an adult. All three birds land on the perch and the chicks immediately begin pestering the adult male for the fish he's holding in his bill. He's unsure who to feed first but settles on one and allows the chick to take the fish off him. The chick swallows it and immediately starts begging for another. The male scans the water below whilst the two chicks peck and click at him repeatedly. He is clearly irritated and tries to move away from them but he's caged in, with a chick either side. He spots a fish and dives down to catch it. As he lands

back on the perch the chicks are all over him and he struggles to find his feet as they fight for the fish. But he is not letting it go until the chick that hasn't yet been fed gets it. Kingfishers are very good at doing this and I have no idea how they know who has been fed and who hasn't. The male bird scans for another fish, locks on to one and catches it. Once again the chicks are on him as he returns to the perch. This time, however, he kills the fish, turns it in his beak and flies off downriver.

The two chicks sit silently on the perch for an hour before another kingfisher turns up. This time it's the female. She feeds one of the excited chicks and then sits quietly for a while, scanning the water for fish. I have been worried about the female for ages, ever since the male left her for the intruding female and the arguing began. She's had a very tough time. She's lost her man and a good chunk of her territory. Watching her sitting with her two chicks pleases me, though, at least she's managed to fight hard enough to keep them alive.

Two weeks later the river is quiet again. The chicks have dispersed from upriver near the nest. I am getting the odd bird sitting on the bridge railing outside the house but they seem to change every few days as the adult male catches up with them and sends them on their way. Life is hard for kingfisher chicks. They have to learn to fish in just a few days and if they don't get good at it quickly they die. The breeding season is not yet over though. The house male and the intruding female have a nest of their own and it won't be long until I hear the clicking again and a whole new set of chicks will be zipping about on the river.

↓
Adult kingfishers seem to feed the chicks in strict rotation and begging doesn't always get results.

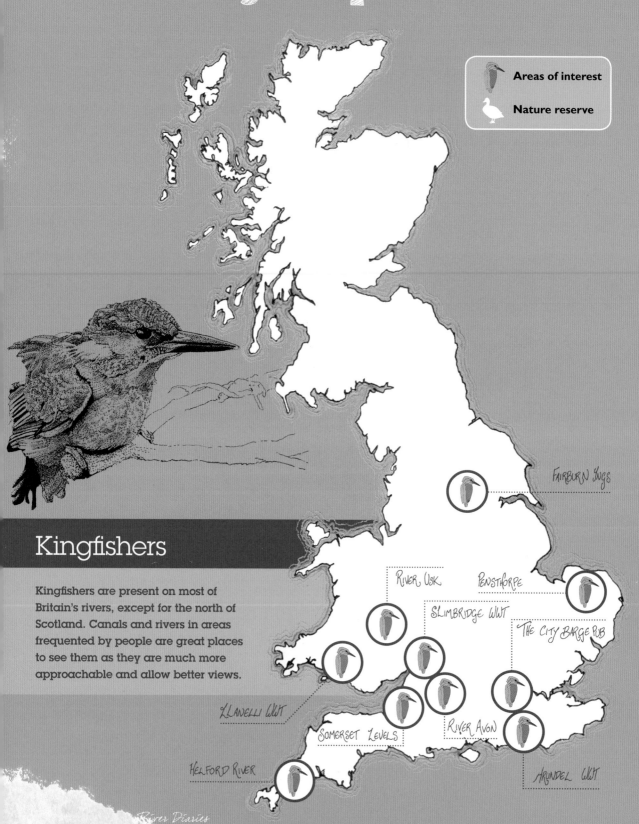

Areas of interest

Nature reserve

Kingfishers

Kingfishers are present on most of Britain's rivers, except for the north of Scotland. Canals and rivers in areas frequented by people are great places to see them as they are much more approachable and allow better views.

FAIRBURN INGS

RIVER USK

PENSTHORPE

SLIMBRIDGE WWT

THE CITY BARGE PUB

LLANELLI WWT

SOMERSET LEVELS

RIVER AVON

HELFORD RIVER

ARUNDEL WWT

Fairburn Yngs, Castleford, West Yorkshire

This RSPB reserve is taking its kingfishers very seriously and has even created a 'kingfisher watch point' next to one of their favourite spots. They are also planning a kingfisher-guided safari in the future which they say will guarantee a sighting.

Penthorpe, Norfolk

Pensthorpe also regards kingfishers as being very important at the moment. They have constructed a large nest bank in the hope of attracting them. It is a large area and there are thought to be several kingfishers frequenting it.

River Usk at Crickhowell, Wales

The Usk is jam-packed with kingfishers. It is also a stunningly beautiful river to walk along. The stretch upriver from Crickhowell is good and has a footpath. There are also other footpaths to be walked near Abergavenny and around Llangynidr.

Slimbridge WWT, Gloucestershire

Slimbridge has a kingfisher hide which is beautifully designed and spacious. It overlooks a pool at the head of a long ditch. On the opposite side of the pool is a kingfisher nest bank and the birds have nested there for years. This allows visitors great views of kingfisher courtship and nesting behaviour in spring and summer.

Llanelli WWT, Wales

Another WWT centre which is great for kingfishers. I have filmed them sitting right next to the hides – so take your camera, there's a very good chance of getting some nice shots.

The City Barge Pub – Strand on the Green, Chiswick, London

The stretch of the Thames between Chiswick and Richmond is great for kingfishers but you have to know what they sound like if you want to see them there. There are also plenty of pubs to watch from. Barn Elms is a WWT centre in Barnes near Chiswick. They have regular sightings of kingfishers.

River Avon, Bath

Bath is great for kingfishers, they can be seen in the city centre and surrounding area. The Mill at Bathhampton is a good spot. The kingfishers are pretty tame and you can enjoy a pint while watching.

Somerset Levels

The levels are absolutely heaving with kingfishers. The huge number of lakes and drainage ditches have created a perfect environment for them. There are hides at Shapwick and Westhay or simply hang out on any of the larger ditches and watch and wait. You'll be hard pushed not to see one!

Arundel WWT, West Sussex

Arundel is a great place to see kingfishers and those who are patient can almost guarantee to be rewarded with a sighting. Even better, you can take an electric boat out on the water and look for them!

Helford River, Cornwall

Like all tidal estuaries in south-west England, the Helford is a great place to spot kingfishers, especially when the tide is out and they're hunting in the fresh water streams at the head of the inlets. Gweek is as good a place as any.

Philippa: After a spring of nest raids and chick murders, the river of death strikes again.

I'm awake and out of bed, one foot on the wooden floor, the other on the rug. The clock says 5.00 a.m. Something is sinking into my brain, part way between dream and reality. I am halfway through a sentence, 'Oh no!' but now I am awake I see Charlie's face, as he walks into the bedroom already fully dressed.

My voice catches up: 'The ducks?'

'He's had them all.'

I run – on the bridge the sound of quacking is still in my head. Concrete on my feet. It is warm already.

It is a beautiful morning, a morning heavy with hangover as if the bucolic heat that we drank our fill of the day before has not yet left, as if the day somehow just carried on drinking after we went to bed, overnight it had never really bothered to sober up or cool down.

What is the point, I suppose: we are at the stage of summer where one day rolls into another beautiful haze of paddling

pools and roses – no school, no cares or concerns. The celebration of the miraculous weather must take over our daily life.

And here was the sun again, warmly drunkenly smiling at me, blissfully ignorant of what has happened, whispering into my ear, 'Hello, ready to party?'

But a pain hits my stomach, the real world butts in and has me leaning against my husband on the bridge and crying.

The fox has eaten all of them, the whole happy combination of characters, the whole set of little relationships. The perfect scene that fills me with joy each time I cross the bridge is gone.

Crying, I need to see it for myself, I need to check because Charlie might have assumed, got it wrong. But there they were. I could already see the big white bodies of the Aylesburys, somehow bigger in death, and other brown bodies, all lifeless. The echoes. All of them gone. I stare and stare, willing them to move but Charlie knows and bows his head. I weep and he holds me and although I desperately want it not to be true I know it is because I heard it – the noise, the quacking, it was what had roused me from my sleep at dawn.

It was still there, the noise, but surely not, surely just an echo? Still sobbing like an idiot, incapable of stopping, in my nightie and barefoot on this beautiful morning I run back to the river. The quacking is a real sound, it is happening, is still insistent like an alarm call, because there in the river below the falls is the baby, the baby of the bunch, the one that I had thought was just going to waste away, she is there and she has a new tribe – a wild tribe.

The fox patrols our garden every night.

Mr and Mrs, the wild mallards, and the other male who has been hanging around are all with her as if they are trying to support her.

Charlie says, 'It's just because she's a female,' but there is no sign that anyone wants to try to mate with her. Only her relentless insistent quacking that something is terribly wrong and that somebody needs to come and do something about it.

I am so pleased to see her and yet so shocked – of all of them I would have thought she would have been the one to go first.

She still has her baby feathers and yet overnight all the creatures that had become her family and social support system are gone. Here she is, having survived where ducks that were twice her size have failed, where guinea fowl that were light and flighty have failed and where chickens that were … well, I suppose the chickens didn't stand a chance.

Her life and her noise rouses me from the state of shock: what are we to do? Instinctively she has chosen the safest place to be, she knows that to be on the water in the middle of the river is the best place and she is staying there. We have no desire to catch her, to freak her out, her own kind seem to be feeling the need to be with her.

I thought of what she had seen, what she had witnessed, weakest member of the team, and I wept again because I couldn't change it, because they all were lost, because although I knew they were just ducks and chickens I have grown to understand that death means you ain't changing it no matter how much you want to. And I weep because this gift of joy which I have given to my children is now taken away and I don't want them to feel the pain of it.

There on the bridge against my husband's frame, in the no-man's time before the clock has woken but after the sun has, I cried till my eyes hurt, until I stopped. Suddenly as the warmth began to hit our heads we only had a desire for coffee and so we left her, noisy but safe, and went back into the sleeping house to talk to the children.

Later in the day I bring home new ducks and place them in the enclosure. I lift the fence and in the river the small brown duck looks as if she has seen a miracle. For the first time since dawn, she stops quacking and leaves the river, running and waddling as fast as she can back to the safety of the enclosure. She avoids me and the lifted fence, however, and I suddenly realise that it was actually her small size that helped her escape the fox. She could leave the enclosure and get to the safety of the river because she could squeeze through the tiny gaps in the fence. The others were trapped.

Although I hadn't been ready to get new ducks so soon, it had been the right thing for her, she snuggled up to the new group as if they were her old friends. After attaching the electric fence to the mains supply instead of a battery, exhausted, I leave them.

Philippa: So often we glance beyond the plants, we don't really see them because we are trying to focus on the wildlife that lives in or above them.

By encouraging the plants to grow, you always provide a home for something else. And in my little patch of golden flag iris in the stream, that is just what has happened this year.

I can't help but notice that the beautiful demoiselles have moved in – that is not an adjective, it is their name.

For a long time, although I appreciated them, I didn't register them as being any different to dragonflies. But now I look I can see there are clear differences. The first I notice is in their flight – dragonflies patrol with purpose, a bit like a spitfire, the demoiselles flutter as if they are on a

To me they are simply jewels, their flights and movements adding extra adornment to my spectacular display of golden flag iris. But like the plants, their agenda does not involve me.

It is the males I am watching and they are exhibiting behaviour quite unlike fairies. My flower display is their territory to be established and then protected with regular patrolling. From the demoiselles' point of view presumably it's more akin to being in *The A-Team* than being a fairy.

I've actually created quite a good area – it has partially submerged vegetation in the water to lay eggs on (very attractive

Like electric-blue fairies they flit and hover.

thread which someone is gently jerking up and down. Next is their size, a demoiselle is slightly smaller.

Like electric-blue fairies they flit, hover and land amongst the green sword-shaped leaves of the golden flag iris. Their vividness forces me to stop as I cross the little humpback bridge over the stream, my eyes fixated, sending my brain into a relaxed trance.

A few weekends ago, I spent some time coppicing hazel on the other side of the bridge. It has exposed the whole bank to sunlight and a couple more beautiful demoiselles have taken up residence.

There are many different types of demoiselle but only two here in the UK – the banded demoiselle, named after its stripy wings; and the beautiful demoiselle, which speaks for itself.

to a potential female), the water itself is not too fast-flowing and so won't wash any eggs away, it has good perching places from where to keep an eye on everything and from where to look out for any females that might appear. Other males come and go, regularly appearing to try to claim a piece of this excellent territory – sometimes fierce fluttering is required to ensure that these intruders don't get carried away.

I settle on the bridge to watch them, fascinated. There is a rhythm to their movements, a pattern followed over and over again. I have seen it in robots – do insect brains and robot control-systems differ much? I wonder. The male spots an intruder, here we go, now we are in for a bit of action, I think a fight is about to commence, there is a lot of flying, and some of it is at close quarters, but it is like

the fight between Colin Firth and Hugh Whatsit in the *Bridget Jones* film before it really gets going. It is pathetic – there is not really any physical contact at all, these guys are fluttering at each other, and my interpretation of this is certainly not *A-Team*. However angry, aggressive and territorial they are feeling, whatever the motive behind it, they look more like beautiful fairy creatures than ever.

Meanwhile, a small brown demoiselle appears and perches on one of the spears of golden flag iris leaves that is bent over. This is the long-awaited female, suddenly she has found her way to the patch, to this most excellent territory. She sits there for a while. The males have fluttered all the way along to the bulrushes and are beginning to return, always close to each other but never touching. How is this a fight?

Although I hate to put thoughts into her head, the female obviously isn't impressed, either and when I look again she has disappeared. In all reality the primitive, robotic nature of her brain just

didn't click into reproductive mating mode because she didn't get the right stimulus, i.e. didn't see a male, but I can't help but laugh – after all, this kind of fighting is frankly foppish.

The males work their way back up the patch blissfully ignorant of the missed opportunity and suddenly, after what must be a particularly threatening flutter, it is all over and the intruder flies off. That told him.

I giggle as I make my way back to the house but realise that this is certainly something that we need to film for the diary.

We spend some time over the next couple of weeks filming the sunny lives of demoiselles and when at last we manage to film them mating it is so revealing and, given my human take on it, moving.

Between them, the showy male and the more dowdy female as they link to mate, make the perfect shape of a love heart, romantic to me, science would say merely practical to them. I wonder.

Slow flowing streams and flag iris are the perfect habitat for demoiselles.

Charlie: **My artificial otter holt, or Charlie's Folly as it is also known, has never really been an object of success and achievement. I am convinced that one day it will work and otters will move into it.**

↓
Building waste takes up more landfill in Britain than household rubbish.

We made it in the spring and I based the design on what I consider otters want in order to feel secure and safe. Of course, everyone disagreed with me, saying that the distance from the tunnel to the chamber was far too long. It is thirty feet from the river to the chamber. My belief was, and still is, that otters would love such a length. The bridge holt at the end of our garden which the otters favour has an equally long tunnel, and I know of other similar holts, so I'm convinced that it is just a matter of time.

I have built several holts in the past, with varying levels of success. The first was in west Wales with some colleagues in the late nineties, with the aim of filming otters for the BBC. It was a 'log-pile' holt which is exactly that, a pile of logs. The otters did move in but we never managed to film them. I then built two artificial holts in my garden; one below the weir, the other above. To my knowledge the one below never saw an otter. The one above the weir was a bit more interesting. In the two years that it had cameras in it I never managed to get a shot of an otter,

I have built several holts in the past with varying levels of success.

although they would spraint outside it regularly. I did manage to film rabbits, badgers, mink, ferrets, foxes, mice, rats and cats in it, though. It wasn't until a couple of years later when I went to inspect it that I discovered the otters had been sleeping in it. The whole roof had collapsed, making the holt much smaller; I always suspected it was too big and spacious and the discovery of a bed area tucked at the back under the collapsed roof proved this.

My latest artifical holt, the one with the long tunnel, avoids all my failings with previous attempts – close to the river, long tunnel, small chamber, second tunnel going out the back, warm bedding inside. My belief is that otters want all these things. The tunnel itself is made of scaffolding planks. I bought a truck-load and screwed them together to form box-section lengths of tunnel. This was not only the cheapest way of making it that I could think of but by far the most environmentally friendly. A concrete tunnel was my first choice but it is costly and polluting. Sourcing my wood from the Bristol Wood Recycling

Project meant that I could buy scaffolding planks for £9 each – so the total cost of the tunnels was around £150 instead of £400.

I got a digger in to dig two trenches. One went from the river, thirty feet back towards the wildlife pond. It then made a ninety-degree bend for a metre or so until it met the big hole which was dug for the chamber. I then dug another tunnel from the chamber to the wildlife pond. Once these were ready it wasn't too difficult dropping the tunnels and the chamber in, screwing them together and covering them up with soil again.

The chamber was the old holt I'd made for my captive otters a few years before and was basically a plywood box a couple of feet square with a lid. They loved the holt so I assumed that the wild ones would too. After everything was in place and covered over, I fitted a camera and lights inside the chamber which were wired back to the house so that I could watch the comings and goings on a monitor.

I'm still waiting, staring at an empty chamber. One day the otters will move in and I'll prove them all wrong! (Them being Philippa and the production crew.)

Building artificial holts seems to be a very popular pastime for many conservation groups. Many of these have worked well (much better than mine). Some are log-pile holts others are far more extravagant affairs requiring more engineering and money. If you want to build your own one I would suggest going online and finding a design rather than following mine.

Otters like a small snug chamber to feel warm and safe.

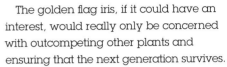

Philippa: It is called either the golden flag iris or the yellow flag *(Iris pseudacorus)* but whatever its name, it is an icon to me.

It signals summertime on the river, a splendid architectural piece, the colour of sunshine, it fills me with joy that the season I have been waiting for for so long has arrived and we have the whole summer ahead.

It, of course, has no idea how wonderful it is, and is really only a response to evolutionary pressures – it is only my set of perceptions and associations that makes it such a beautiful thing.

The golden flag iris, if it could have an interest, would really only be concerned with outcompeting other plants and ensuring that the next generation survives.

It has evolved to exploit the margins and has become a plant that has a competitive edge on others which allows it to live at the edges of streams and rivers.

It is quite happy to have its feet in the water on account of special adaptations. And now I must confess I did my degree in ecology and conservation at Birkbeck many years ago, but still ringing in my ears is the voice of my flowering-plant lecturer, a man who is immersed so deeply in the complexities, anatomy and evolution of flowering plants that it is difficult to imagine him knowing about the mundanities of life. 'Don't forget the aerenchyma,' he would say, smiling, with a wag of his finger, and we would smile back.

Traditionally people have other more useful pieces of advice echoing around the caverns of their brains, like their times tables or 'Don't forget to wear your vest' but for me, 'Don't forget the aerenchyma', has stuck.

And perhaps it was that, or the knowledge of this native plant's value in the watery ecosystem, or indeed just because I think that they are beautiful, which made me devote a small section of our little stream to them. Out went the collection of conifers in different colours that had been planted by my predecessor and in came a collection of rhizomes.

Muddy and deeply unattractive but crucial to the survival of the flag iris, inside this sludge-coloured, slimy bulbous root is to be found the 'aerenchyma.' They are air passages, or more specifically intercellular air spaces. These passages enable this plant to survive under the water and in the mud, where lack of oxygen would kill others. All through the winter little else exists, the rhizomes act as a food store and these spongy tissues within hold the air. Then in the spring they sprout spear-like green leaves which grow by the day until finally in June they burst into flower.

The flower is another means by which the plant survives and regenerates, it just so happens that it looks a lot nicer than a rhizome. This flower is not designed for

Flag iris like the reedbed opposite our house – it is perfect for them.

This flower is not designed for us, however, it is designed for insects.

us, however, it is designed for insects; pollinators. Every feature, from its architecture and size to the yellow colour and other adaptations invisible to us. The iris group have markings, guides if you like, for bees to follow, in just the same way that a landing strip at an airport is painted with white lines or laid out with lights. Some are even ultraviolet, and so invisible to humans, but perfectly visible to the bee. And so it is that I watch bumble bees and insects visit the flower, landing on the platform of the lower yellow sepals and following the invitation on down the tube to collect the pollen, just as the golden flag iris and evolution intended.

None of it is for my benefit and yet it inspires and uplifts me to see, now, three years after that first planting that there is a huge display of flag iris up and down the stream. The rhizomes have been spreading under the water.

I wonder if actually there is some value in my appreciation of the flower, some evolutionary benefit to the plan. After all, if it wasn't for me and my love of the golden flag iris there wouldn't be any here.

Philippa: It's all very well getting to know the animals above the water, but the river is home to so many more below the surface. This is where the creatures are endlessly fascinating: dragons and demons in the form of hundreds of different nymphs and crayfish monsters, with their morphing bodies that defy imagination and inspire horror films.

We can't ignore this part of the habitat, it is critical in providing many of the bottom layers of the ecosystem. The picture of a river is only partly complete if we don't submerge ourselves. That said, getting under the water is not easy, especially if we are hoping to watch small things over a long period of time. So this weekend we have decided to clean out the old aquarium, create in it a river bed and fill it with creatures making a habitat where they can interact naturally so that we can observe and film them.

The large tank sitting idle in the garden is fibreglass and has glass windows, it was built as a play place for an otter that we once looked after and is perfect for what we need. It is full of green smelly water and an old scaffold tower and will take most of the weekend to clean out.

Charlie is brave enough to try the siphon method to empty it and sucks on a large piece of drainage pipe. There is a gurgling noise but little else, not even a stirring in the green waters. He sucks again, nothing, and again, he can only just get his mouth around the nozzle. He goes to suck it again and is looking up to tell me that he thinks the pipe is too big when it starts to spew the stinking, slimy water and he has only narrowly avoided a mouthful of it.

The kids soon get bored and end up singing, dancing, swinging and fighting. Next day we clean and refill it. We carefully place large rocks and native plants, gravel and stones to mimic the river bed, hiding places for mayfly larvae, reeds for dragonfly larvae to climb and large stones for bullheads.

Finally it is time to go fishing for tiddlers and after a welly-filling expedition upriver with some jam jars, a frustrated Dad and a few over-enthusiastic boys, we are filling the clear clean aquarium with sticklebacks, bullheads, minnows, larvae and tadpoles. We duck down to the window to watch them settle and all agree that it is worth the effort.

The tadpoles seem to disappear quite quickly but over the next week the

Most fish have no parenting skills, but the aquarium is already revealing that the stickleback is very different.

stickleback fascinates us most. He seems to have chosen a place near the glass window to make his nest and so he is easy to watch. We christen him 'Frank' because, like all male sticklebacks, he has bright blue circles around his eyes and his jaw is red. There have been studies which suggest that, for the benefit of females, the stronger the intensity of these blue eyes and the contrast with the red jaw, the more superior is the male stickleback.

In stickleback world the males are in charge of almost everything in the reproductive process. He builds the nest in the crevice of an old log, gluing bits of vegetation together with sticky thread apparently coming from his kidneys. His nest has got to be impressive enough to

tempt a female in. He isn't interested in a relationship and only wants her eggs, from then on he's a single parent. It might seem against all odds to us but Old Blue Eyes isn't daunted, even when Gus knocks the top of the log by mistake, destroying the tiny nest underneath and he has to build another one.

In a couple of weeks with some fancy dancing and bright eye flashing and he has managed to seduce a female back to his nest. We have managed to capture most of it on film and we watch enthralled at the care he takes of his new eggs. He even fans water over them to ensure that they get waves of extra oxygen. Most fish have no parenting skills, but the aquarium is already revealing that the stickleback is very different. Frank really cares.

Diary

Charlie: I've spent a huge chunk of my life trying to get the perfect shot of a kingfisher dive. I have failed so far but I have learnt a lot about kingfishers diving. The actual sequence from leaving the perch, hitting the water, grabbing the fish, and returning to the perch takes barely two seconds but slow the action down and you see an incredible degree of skill and design.

↓

Kingfishers are very choosy about the fish they eat – minnows seem to be favoured.

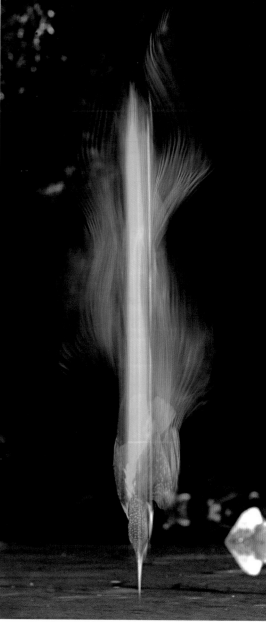

Firstly the kingfisher has to choose its fish. The birds are very particular about what they eat. The kingfishers on our river have a strict hierarchy, with minnows at the top, sticklebacks and bullheads as a last resort. If they are catching food for their chicks or for courtship feeding, they'll catch just about anything, even small crayfish.

Once the kingfisher has picked its target it locks on to it, works out its angles, draws its feathers in tight and dives thin and bullet-like. On the way down it works out the refraction of the water (water bends light and so things aren't actually where they appear to be when you look at them in water), it calculates the depth of the fish, flaps a couple of times for increased speed, shuts its eyes to protect them, opens its beak and hits the water. As the water engulfs it, it turns silver as air is trapped

around its feathers. When it reaches its calculated depth its beak snaps shut and grabs the fish. The kingfisher then slams the brakes on by extending its wings out. It flaps them once, throwing it into reverse, reaches the water surface and with a mighty effort flaps clear and back up to its perch. The poor fish is then battered head first against the perch.

The hit rate of kingfishers is pretty high, I would estimate about 70 per cent when the birds are fishing in water less than a foot deep. The hit rate does decrease, however, as the water becomes deeper, not just because the bird has a more complex set of problems to deal with, but because the fish have a slightly higher chance of reacting to the kingfisher as it bears down on them.

I have never actually counted how many fish a kingfisher eats per day and I would suggest that in wild conditions it would be impossible. I would estimate from watching for so many years that an adult bird would eat around fifteen to twenty fish a day, depending on good conditions such as clear water. I currently have a kingfisher with a broken wing and I feed him about ten fish a day. The vet says he's a little fat but he's not burning much energy so it's not surprising. If he was out in the wild his food intake would be substantially higher.

Kingfishers do occasionally make mistakes when diving. I once read a story of a kingfisher who drowned diving into a jar of tiddlers some kids had caught. I have seen kingfishers catch stones from the bottom of the river. I even once saw one catch a large leaf by spearing it. It then sat on its perch for a while trying to get the leaf off its beak. I guess when you're performing something so fast and so spectacular with your eyes shut, you can be forgiven for making the odd mistake!

Philippa: 'Mum, I have a question.' 'Yes, my angel?' I smile beatifically up at Dominic the botanist, from the charity Plantlife, who has come to study our flood meadow, placing my hand in a Madonna-like pose on Gus's blond (if slightly grubby) hair.

The sun shines in the flowering meadow, we are warmed by it and so are the insects that are coming alive around us. I wait for the intelligent question from the mouth of my cherub, a carefully planned response from my thoughtful middle-child about the tiny St John's Wort flower, or the other flowering species Dominic and I have just been showing him.

'Can I go and play on the Wii now?'

I open my mouth, but no sound comes out, my Madonna moment explodes and floats off in pieces.

Luckily Dominic is a father. 'Well,' he says, 'to your credit you did wait for about ten minutes while we waffled about plants, before you asked, that was very polite.' Gus thought that Dominic was cool when he taught him how to make firing weapons out of plantain. Now, sensing an ally, he instantly likes him more and looks up at me, eyebrows raised like a spaniel puppy hoping for a titbit.

'No,' say I grumpily, and he strops off back to the house, head down, in a sulk.

It does, however, leave us free to continue ferreting around in the long grasses getting excited by hairy leaves and tiny flowers scattered in the vegetation, spilt sparklers.

'I'm sorry.' I feel I have to apologise, but also I *am* sorry – the sad fact is that although we all assume kids love nothing better than having the freedom to run around outside, they would prefer, most of the time, to be inside in front of a computer.

A muddle of meadows

Flood meadows are not water meadows
Water meadows have been created by humans from a system of channels and sluices leading off a river. Fertile conditions are recreated by artificially flooding a piece of land before using it to graze cattle or sheep or make hay. Flood meadows, however, are entirely natural and wet most of the time.

Flood meadows are not wildflower meadows
These are on drier ground which is not as nutrient-rich and so home to a different set of species.

So rather than being the perfect mother I want to be, I turn into the demon from hell because I am insisting that they stay outside and explore the natural world. I am the one always doing battle with the screen, I am the one saying 'No, no, no.'

I am also saying, 'Yes, yes, yes', I spend my life facilitating what I think the kids should want. We live in the countryside, and we struggle to afford a place with a massive garden and lots of wildlife because I think it should be an important part of their childhood and I want them to run and have space to cycle, play, build dens and explore.

I sometimes wonder if it is worth all the hard work. Is what I want a nostalgia for a type of childhood that is in the past, and not relevant to today's generation? Are they better off learning their way around computer systems? Is that what they will really need in their future?

I kneel and the grasses are above my shoulders, butterflies dance all around me, insects make a noise and there are flowers everywhere. In fact the closer you look into the vegetation the more you see. I can't help but listen to my heart insisting that this is the real world. This is their home.

Dominic has already been distracted and moved on – he has very kindly come to help me with a species count and management plan for our flood meadow.

I have discovered that the flood meadow is indeed a precious habitat. Although it is man-made to a degree, the high biodiversity comes from a traditional twice-yearly mowing and clearing regime once in July and once in September. Flood-meadow species are adapted to having their feet in water. The meadows are traditionally damp and spongy underfoot and are often inundated. These species are quite particular about their

conditions and therefore unable to live in other habitats. This makes them vulnerable if we lose our flood meadows. Flood-meadow species are also used to the regular inundation and the resulting silt deposition which adds to the fertility of the soil. And so in recent years the naturally-occurring wildflowers have been cleared to make way for agriculture or development.

They are also exposed to changes upriver, difficult to protect and control. If a farmer further up fertilises fields or uses pesticides which run off into the river and flood the meadow then there is very little anyone can do about it. Too much nitrogen in the water, for example, will alter the balance of wildlife and allow a more vigorous species to take over.

Here, however, this little patch of flood plain has been spared because of its proximity to the house on the other side of the river, and as a consequence Dominic regards it as ancient flood meadow, a safety deposit seed bank.

So what do we have here in our miniature ancient flood meadow? The answer is fascinating and proves just how important the correct management regime is.

↓

Ancient flood meadow is rare in Britain. We're very lucky to have a tiny bit of it.

There is a fence through the middle of the meadow – the side nearer the house is where the children play on their trampoline and see-saw. It is an extension of the lawn and is cut at the end of every summer so that we can keep it under control.

The other side of the fence is just a paddock, we have no practical use for it and so no need to mow it regularly, it just exists. Charlie often says 'leave nature to itself and it will be fine' and I think that most people assume that to be the case. However this paddock proves that it isn't necessarily true.

Sometimes even the most natural spaces need managing for the maximum biodiversity. The wild world is a dynamic place, constantly changing – plants grow and as they change shape they alter the local habitat. Often a natural place will slowly but surely change, for example, into a woodland. This process is called succession and there are clear stages.

At each successional stage different creatures live in the habitat. In wetlands the clear ground makes way for hazel, alder and willow which then grow into wet woodland. Slowly they dry out the soil and allow other tree species to take root. The habitat changes again from wet and open to drier and gradually more shady. The other plants and insects which live there will alter too – smaller flowering plants might get shaded out, butterflies will move away to sunnier places. In many decades, if left alone, the wetland may dry completely and turn into the last successional stage in the UK: the 'climax community' which would be ultimately oak woodland. Then that, of course, will change again if there is a fire or a clearing created where a tree falls. Suddenly a patch is exposed, allowing sunlight to reach the forest floor. As if from nowhere sun-loving flowers spring up, but actually their seeds have been lying dormant on the forest floor all this time, waiting for the opportunity to grow.

When we think about conservation and management we are hoping to maximise biodiversity and allow habitats to exist for the maximum number of species so this means maximising the successional stages. It seems odd that in the name of conservation we are interfering in the natural processes at work, in some instances holding back the natural process of succession. This is particularly the case with wetlands and river environments which are more dynamic habitats than most.

I have been able to observe this in our own back garden already – the wildlife pond which we created just a few years ago has changed, reeds have grown into thick banks, the water plants are crowding out the open water spaces. On the bank, small alder and willow trees are growing

↓

Yorkshire fog.

and, if I let it, this would all affect the delicate balance of the flood meadow. In fact the paddock part of the flood meadow has already gone into decline, my lack of management and the fact that I haven't bothered to mow it has turned it into what Dominic describes as rank grassland.

We do a species count, taking a square metre of ground between us in the paddock and counting the number of different plant types we find there. There are two. Wild angelica, a large and bulbous plant and hard rush, both typical of marshy conditions.

The dense mat of fallen long-dead grass means that few of the smaller species can grow there, it is just a thatch of rotting material. I try but it is hard to even pull it out by hand.

So we hop over the fence and into the 'garden', the area which is mowed once a year. The difference is remarkable – we find eleven species in just one square metre and thirty-three overall. They are fascinating plants, an equal mixture of wildflowers grasses, sedges and rushes.

Dominic points out the jointed rush (a great wetland indicator), the false fox sedge, a beautiful delicate grass called Timothy-grass, sweet vernal grass, the wonderful Yorkshire fog and, lastly, the hairy sedge, with tiny hairs all along its blades backlit in the sunshine. I had no idea there was such variety.

We crouch down and peer at the purple bush vetch and yellow meadow vetchling, such tiny plants I wonder how they can possibly compete with all these grasses for their fair share of sunshine. Dominic explains how they use tendrils to grasp the taller plants for support so that they can get up high.

I had no idea we have St John's Wort, tiny yellow star-shaped flowers, named because they were harvested around now – St John's Day, the 24th June.

We are so busy with our head in the flowers that we don't notice that Gus has returned, creeping up behind us and firing us with plantain seed heads, then running off and laughing. His brothers join him and they go crazy in the meadow – firing, chasing, battling, rolling. It makes me smile – OK, he may not be ready for the detail of each plant but on some level something is going in.

As the warm glow of the sun retreats behind the riverbank trees and leaves the meadow in shade, Dominic and I share a pot of tea and make a plan. Depressingly it involves a lot of work … but I don't have to think about it until September.

Many wildflowers like bush vetch (top) have adapted to compete against tall grasses.

Charlie: Witnessing the emergence of a dragonfly from larva into its fully-winged adult form is, to me, one of the greatest wonders of the natural world. It represents the most staggering achievement in natural design and engineering and is one of the few things that I have seen which leaves me unable to comprehend how evolution made that jump.

The best thing about it is that it happens all around us in the spring and summer, in any lake or garden pond. Of course, like many things in nature it doesn't really happen on our timescale; some parts are fast, others are painfully slow. However, armed with a camera, it is possible to speed the process up and take a look at it in a wholly different way.

I made several attempts to film dragonflies hatching but failed to get the perfect shot. My first attempt was a disaster. I had assumed that the dragonfly would take several hours to hatch out of its larval skin; it actually only took a few minutes. As I was attempting to shoot a timelapse of the event, all my settings were way out. My aim was to speed up time a little so that the minutes it took the dragonfly to emerge only lasted for seconds – around fifteen to be exact.

The problem was that I got my calculations wrong so I ended up speeding up time a little too much – the whole event lasted about two seconds! – a bit quick for the average viewer to see.

So I adjusted my settings and waited a few days for the next larvae to emerge. I was attempting to film larger dragonflies such as southern hawkers and emperors, which seemed to do things in a similar way. Firstly, they tended to hatch at or just after dusk, which made lighting them a little tricky – I was bound to using flash guns. Secondly, they were fairly alert throughout the process so I had to sneak around them whilst they were hatching so as not to disturb them. However, I did learn, after a few attempts, when they were going to hatch and this proved very useful in getting the shots I did finally achieve.

Dragonfly larvae can live underwater for up to two years. They are powerful and voracious hunters at this stage of their lives, able to catch a whole array of different species even including small fish! They have an amazing and powerful set of jaws and grabbers, called the labium, which can extend out to stab and grab passing meals, before retracting to devour the prey. These weapons are amazing to watch in action and make the dragonfly larvae look like some kind of alien; indeed the mouthparts of the aliens in the film *Alien* were based on them!

Dragonfly larvae tend to start emerging in the spring and continue through the summer, stopping around July and August.

They start the process in several ways – the southern hawkers and emperors that I was filming tended to begin by clinging to an iris stem, which they then spent at least twenty-four hours doing with only their heads sticking out. I had placed several iris stems in their specially constructed tank so that I could limit and control what was happening. This allowed me to be ready for them with cameras and lights when the action started.

The next stage would happen at dusk. The larva would begin to climb the stem. When it reached a certain height, generally around eighteen inches off the water, it would stop. It would then remain in the same position motionless for up to two hours. During this time I would fiddle around with my cameras and try to get everything working properly together, ready for the moment of hatching. What I noticed, though, was that a few minutes before hatching the larva would start thrashing its tail around. I put this down

to it loosening itself inside its skin, in order that it could hatch from old skin into new. But the scientific understanding is that the dragonfly is making sure there is nothing around it that is going to get in its way when it emerges – this does make sense.

After the thrashing around bit stops the larva then goes motionless and holds on. A few minutes later the thorax starts to split as the dragonfly begins to heave itself out of its old skin. Once this process has started it tends to proceed fairly quickly. First the thorax and head emerge and then the top of the abdomen. At this point the dragonfly begins to arch backwards. As more of it emerges the dragonfly begins to hang upside down next to its skin until it is only being held in place by the end of its abdomen. It can hang like this for up to an hour until suddenly it makes a flip, heaves itself up and pulls the last bit of its tail out of its old skin. It then clutches the old skin and allows its new wings to hang down.

Dragonfly larvae are hard to spot, they hunt by ambush.

It was around this stage that a lot of my dragonflies died.

Just after hatching, the dragonfly is incredibly soft and fragile – not like the tough-skinned bully that it will become. Its wings are folded like a parachute and its abdomen is so weak and soft that the slightest touch will rupture it. The dragonfly needs to allow itself to dry out and let its skin toughen up. The problem is that it can't help moving around and adjusting its grip on its old skin. This often leads to disaster. The first dragonfly I successfully filmed hatching killed itself by spearing its abdomen on a cut iris stem. I had filmed it emerging and left the camera snapping away on its wings while I went inside to watch TV (it was about 11 p.m.). When I came back out an hour later the dragonfly was gone. I couldn't understand it. I searched about and eventually found it almost cut in half on an iris stem. It had obviously let go by mistake and fallen the few inches onto the stem, which had cut right through it. I felt incredibly sad as it flailed around helplessly and felt the best I could do would be to stamp on it. I was quite distressed about this. I had watched it hatch, witnessed its struggle and marvelled at its incredible engineering; to see it flailing around half-dead and useless was bitterly sad.

This happened on two other occasions. One dragonfly kept falling off the reed and I tried desperately each time to rehang it but to no avail. It suffered the same fate when fluid began leaking from its wings and I realised it was useless trying to save it. I spoke with dragonfly guru Dave Smallshire about it, fearing that I was doing something wrong but he explained that it was a common occurrence. I guess such engineering might be incredible but so often in nature, it comes at a price.

I did have successes though. One dragonfly I filmed did everything just perfectly. After hatching and righting itself on the old skin it then sat motionless all night and I managed to film its wings as they uncrumpled and expanded. Once fully expanded the wings then dried and went brittle. In the morning I came out to film the dragonfly who looked immaculate in the morning light. I got a few shots of it, close-ups of wings and head before it suddenly lifted off the reed it clung to and flew off over the house. I was stunned, not only had it emerged from its pupa in a staggering feet of engineering, but it could fly with incredible skill and manoeuvrability immediately, without any lessons.

Photographing dragonflies

Charlie: I love photographing dragonflies. Firstly they are easy to find and approach, and secondly they are a taste of the exotic in our own backyards. With the right technique you can get some world-class results.

My first efforts to photograph dragonflies flying were a disaster. I stood by my pond with a medium-length zoom lens and a camera armed with a flashgun and attempted to snap some winners. After half an hour or so of machine-gunning with my camera all I ended up with was several hundred out of focus and badly exposed blurs.

My technique seemed sound to me – fast ISO, fast shutter speed (around 2000th of a second) and a fairly swift set of arms and eyes. Clearly none of these were fast enough. My biggest problem was focus. I was relying on the auto focus to try and keep up with the dragonflies as they zipped past me. When a subject is moving so fast and being followed so erratically the auto focus simply can't work out what you want it to focus on, and with a medium-length zoom lens the dragonflies were not very big in frame which confused it even more.

Field assistant Ian was having some luck with his 400mm lens. He'd got a few decent shots of the dragonflies. So I thought I would give it a go. I proudly arrived at the pond with my 500mm lens. The difference in size between Ian's 400mm and my great big 500mm was significant but after standing around for another half an hour I got nothing but an aching arm and I left empty-handed.

My next attempts were more successful and not for obvious technical reasons. The reason for the better results was simply time. Rather than marauding about the pond chasing dragonflies I simply sat and watched them. There were two species around – the little red common darter and the larger

more impressive southern hawker. What I noticed by sitting and watching was that most of the dragonflies kept to a certain flight pattern; I assume they were hunting and patrolling their own territory and they seemed very focused on taking the same routes everywhere or flying in the same circles. If I sat still these flight paths would take them very close to my face and, what's more, they would hover close to examine me.

This was the breakthrough I needed. If I could just stop the dragonflies moving so fast I could focus on them and get some shots. So I stopped using the long lenses and opted for a shorter 100mm lens. This would not give me a massive close-up unless the dragonfly was really near by but that's what I wanted. It wasn't long before it worked. A common darter came very close and hovered in front of me. I flicked my lens to manual focus and had just enough time to rattle off a couple of shots before the dragonfly zipped off on its circuit. I checked the LCD screen at the back of my camera and was surprised to finally see a pin-sharp picture of a dragonfly. Getting rid of all the fancy kit had worked – I was using just a lens and a camera – no flash, no telephoto, no auto focus. I had set the ISO to 400 and the aperture to f5.6, which was allowing me to shoot at around 2000th of a second – just about fast enough to freeze the bodies of the dragonflies but still retain a bit of movement in the wings. Freezing the subject completely often makes the image look false or just lacking in life and energy, a bit of wing blur makes the shot look much more real.

I spent the rest of the day getting more of the same but each time honing my skills. By the end I was trying to get on the same level as the dragonflies by lying face down in the mud. The shots worked and I was very pleased with them.

> What I noticed by sitting and watching was that most of the dragonflies kept to a certain flight pattern; I assume they were hunting and patrolling their territory.

Philippa: Deep in the aquarium Frank is having problems with the Mob.

After she laid her eggs the female left. He never saw her again but he didn't care. He was only interested in having a family and now he is a determined 24/7 single dad, surrounded by a mob of hungry minnows desperate to eat his brood.

As soon as they come home from school the children abandon their collection of snack boxes, rucksacks and works of art and rush over the long grass to the aquarium, diving under the heavy black fabric we have pinned up to stop the light reflecting off the glass. Underneath, it is like a different world, and they are hushed – here is real widescreen drama.

On the left a mob of minnows have massed. On the right swims Frank and behind him hundreds of tiny blue-eyed baby Franks. The odds are completely stacked against him but again and again he dives forward, aggressively defending his cloud of babies.

Again and again a minnow tries to sneak by and steal one and he darts forward and attacks it.

The boys can't get enough of it.

'Look, the babies are so cute.'

On the left, the Mob ...

'It's like *Star Wars*.'

'Fish wars, look at that fighting.'

For Old Blue Eyes there is no question of giving up even under such ridiculous odds. We leave him after a while for tea and homework, and I wonder how he can possibly defend those babies like that for 24 hours a day. But again I seem to be judging by human standards, this fish is superhuman. I shouldn't have been at all surprised that he kept most of his babies alive and over the next few weeks we watch them grow. A few are indeed claimed by minnows, young trout and even dragonfly larvae, but now there are other stories unfolding in the aquarium.

... on the right, Old Blue Eyes.

Charlie: At the top end of our garden is a bridge with an old drainage pipe at the base of it. During the dry months this old pipe is regularly used by otters to sleep in during the day.

The hardest part of filming otters is finding them. If you can narrow down their location then you stand a much better chance of filming them. As a result the bridge holt is the perfect place to work. By simply putting a smooth layer of sand outside the drainpipe I can tell when there are otters in residence. As I have been filming the comings and goings of otters in the bridge holt for several years now I have got my timings down to a fine art. I never have to wait for more than an hour as they always come out around dusk. So when Gus announced that he wanted to go filming otters, off we went to the drainpipe.

up two infra-red cameras and four infra-red lights. As the bridge has a road going over it I can do all this from the car. I'm in a hurry though, darkness is beginning to fall and I need to move quickly. The cameras are ready. I quietly sneak back to the car to check that the pictures are working on the monitors that I have on the dashboard. Everything looks good – a rare treat in the world of rigging where something always goes wrong.

It is almost dark. I run back to the house, grab Gus and run him back up to the car. I quietly open the passenger door and put him on the seat. 'Here are the monitors,' I whisper, 'watch them.'

Poking out of the drainpipe, large as life, is the head of an adult otter.

● REC

Tonight is the night. Gus is back at the house waiting for me to pick him up. He has his pyjamas and a dressing gown ready. I'm rigging kit. In order to film the otters coming out of the holt I have to set

For the record, our second son, Gus, is an interesting fellow. He has more focus on what he is doing than any child I've ever met. However, this is only true if he is very interested. My sons love the idea of staying up late and going out filming wildlife with dad but they soon realise the reality is not nearly as much fun as the idea and we usually end in boredom and failure.

The last time I tried to get Gus to see an otter we were in an almost identical situation. He lasted the grand total of nine minutes. I guess, aged five, nine minutes is an awfully long time to do nothing. Of course, we didn't see anything, never mind an otter.

This time I'm hoping Gus can last a little longer. He watches the monitors while I pack things into the boot and get a spare battery. 'Dad, there's an otter,' Gus whispers. I ignore him. I assume the few seconds he's watching the monitors he is already bored and he's decided to wind me up. 'Dad, there's an otter,' he whispers again. 'It's looking at me.' I suddenly realise that he might actually not be bored and fibbing. I run round to the driver's door and look at the monitor. Poking out of the drainpipe, large as life, is the head of an adult otter looking straight at the camera. I fumble and hit Record on the two monitors. It looks like a female. After a moment she goes back into the tunnel. 'Well done, Gus!' He's not bothered. 'Now can I play a game on your phone?' he asks. 'No, watch for the otters.' Of course Gus ignores me, being the hero of the moment, and takes my phone from by the gear stick and starts playing a game. I watch the drainpipe on the monitors intently. Within ten minutes the otter's head appears again and then another! Mum and cub. 'Gus, look!' I whisper excitedly. Gus glances quickly at the screen and sees the otter, then immediately carries on playing on the phone. I take it away.

'Gus, look, there are two otters, isn't that exciting?' 'Dad, I've already seen them.'

Gus and I (mainly I) watch the otters as they both squeeze themselves out of the tunnel and begin sniffing around on the sand outside the holt. The mother finds a spot to spraint and the cub goes down to the water's edge for a drink. I can see how big the cub is. It is well developed and not much smaller than the mother. I'm not sure whether it is a male or a female though. The mother pushes past the cub and enters the water, the cub follows.

I watch to find out which direction they go; they vanish into the murk. I think they've gone upriver. I hope they have.

I phone Philippa and tell her to watch out of the window to see if the otters have actually gone downriver. If they have, she'll see them as they pass the house. 'Right, Gus, time for bed,' I say. 'Can't I come with you?' My plan is to drop Gus off then drive one mile up the river. I need to get well ahead of the otters so that I can set up to film them before they arrive.

Despite Gus's protest I drop him off, swap some kit around and head off. I've got Richard (director) and Ian (camera assistant) with me and we're going to try and get some really close shots of the otters tonight.

We get to the Manor. It's pitch dark. We get the kit out of the car. I have my

↓
Mum and cub leave the drainpipe holt for a night's fishing.

mobile unit with me – a big tripod, camera and recorder on it with big lens and powerful red hunting lights. We make out the river at the end of the Manor garden. This stretch is about two hundred metres long, dead straight and shallow, giving us the best chance of seeing the otters and getting some shots of them long before they reach us.

I find a spot on a shingle bank. It's perfect, nice and low with a great view down the long straight stretch. Richard and Ian get settled on the bank to my right. I have a dead trout in my bag. I want to know whether the otters will take it if I put it in the river. I tie it to a rock and place it in the river just below the surface. Then we settle and wait.

Two hours later we're freezing cold. It's gone eleven and all we've seen is a rat. My red lights are powered by two 12v batteries, they have been on for a couple of hours now and sooner or later the batteries will fade. I don't really understand why the otters are taking so long. The distance between the bridge holt and our location is only about a mile. My only concern is whether they travelled that distance very quickly and

we've missed them. I keep watching down the long straight, hoping.

I've got something! The tiniest glint of eye-shine. I see it again, way down the river. Then full beam; two bright eyes pop up out of the murk and move about in the blackness. Another set of eyes pops up just behind them. 'There they are,' I whisper excitedly to Richard and Ian. I swing my lights fully on to them. They are still a good two hundred metres away but the eye-shine is very bright.

I power up the camera and recorder and try to find the otters through the viewfinder; I pick them up crossing from the far side of the river; they appear to be hunting in a shallow pool but they're heading our way. Without warning my infra-red camera and one of my lights goes down. I'm left suddenly looking at an empty blue screen. I panic. I press buttons and fiddle with wires but nothing. I grab hold of my rucksack, open it and yank all the wires and 12v batteries out of it. I fiddle with the connections and to my relief the light comes back on and the picture snaps back onto my viewfinder. Just in time, the otters are heading our way rapidly. I pick them up again and get shots of their bright

Realising that otters would take a dead trout was a big discovery for me.

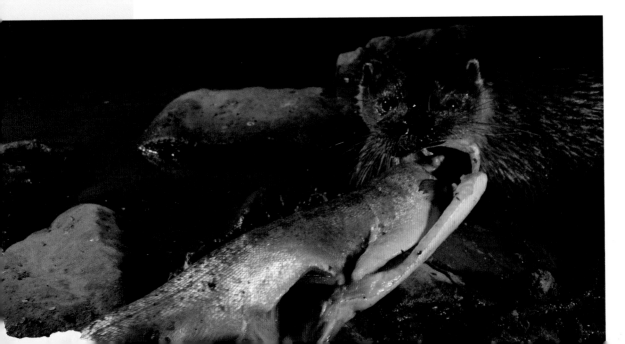

eyes heading towards me. One of them is ahead of the other and moves much faster up the edge of the opposite bank. I manage to get some shots as it begins nosing around in a small set of rapids. The other joins it and they move towards us. When they are about thirty feet away they stop together in the water. 'Have they smelt us?' I zoom in to get shots of their heads. The smaller of the two, the cub, moves off from the mother and grabs the dead fish; the mother piles in too but the cub has it firmly and won't let her get it. She noses around it as the cub tries ripping it off the brick I've tied it to. It won't break free though and the cub drags both fish and rock into the middle of the river before eventually ripping the fish off the brick. The mother goes in again to get some of the fish but the cub turns and swims off downriver fast, waving his tail as he goes. I've seen giant otters in Peru defend their catches by waving their tails around but never one in Britain. I am very impressed with the cub.

The mother is clearly not. She sniffs about where the fish was for a moment then heads up to us. She is mid-channel when she reaches the bend where we're sitting. I film her as she rounds the bend, checking us out. She dives and I lose her for a second, then spot her again as she pops up under some roots to sniff us. Satisfied that we're nothing too much to worry about, she gets out on the bank opposite and sniffs around on the shingle beach. I try to keep track of her through my lens but don't do very well. She inspects the entrance to an old disused holt, spraints next to it and then enters the water again right opposite us. She checks us out the whole time, as if she knows something is there. I pick up a few more shots of her as she swims off round the bend and walks up the shallow rapids before I lose her to the river.

The cub came so close I could barely focus on it.

Richard and Ian can't contain themselves. I'm lucky to have spent thousands of hours filming otters and I've had some amazing encounters, for Richard and Ian though an encounter like this is a first so they are delighted. We chat excitedly to each other while I scan the river for the cub. We wait ten minutes and when it appears it's moving fast up the middle of the river towards us. I try to keep track of it with my lens but lose it. A second later ripples emerge from the vegetation a few feet to my right and suddenly the otter appears on the surface, swimming towards my feet. I tilt the camera right down to pick it up and find it just as it edges toward me. The otter stops and gets out of the water to sniff my boot. I freeze but keep filming. The otter gets back in the water, swims round to my other boot and has a sniff of that. I can't believe what's happening. This young otter is just inches from my foot and I'm sitting there with a massive camera and lens with nearly two million candlepower of light shining at it from three feet away. Completely unfazed and relaxed, the otter heads off upriver. I keep track of it and keep myself together until it has gone from view; then Richard, Ian and I fall about laughing. Richard has got the whole lot on tape and is ecstatic. Even with my hours of otter watching behind me I cannot believe what has just happened.

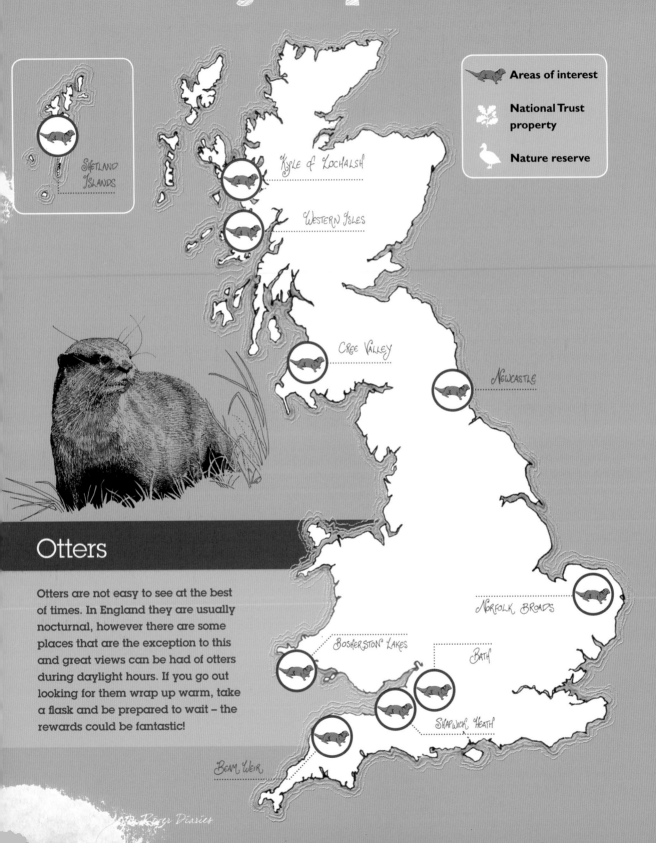

Ten great places to see ...

Areas of interest

National Trust property

Nature reserve

Shetland Islands

Kyle of Lochalsh

Western Isles

Cree Valley

Newcastle

Otters

Otters are not easy to see at the best of times. In England they are usually nocturnal, however there are some places that are the exception to this and great views can be had of otters during daylight hours. If you go out looking for them wrap up warm, take a flask and be prepared to wait – the rewards could be fantastic!

Norfolk Broads

Bosherston Lakes

Bath

Shapwick Heath

Beam Weir

Shetland Islands

Just about anywhere in Shetland is good for otters and they come out during the day. The ferry terminal at Toft is a particularly good spot.

Kyle of Lochalsh, Highland, Scotland

The otters at Kyle wait for the fishing boats to come in at night before boarding them and eating the scraps. Spotting them is hit and miss, but when they are about, otter watching doesn't get any better.

Western Isles

Mull and Skye are particularly good for otters and more accessible than Shetland. Otters are active throughout the day. The best time to look is when the tide is going out or coming in, but not full.

Cree Valley, Dumfries & Galloway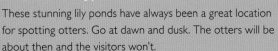

This area of stunning woodland still boasts a healthy population of red squirrels. The River Cree is well known for its otters and they can sometimes be seen from a specially-built observation platform. Follow the trail from the Wood of Cree to the platform that sits on the edge of Loch Cree.

Newcastle & Gateshead

Otters do very well on the Tyne – they can be seen during the day, but you'll have more luck at dawn and dusk. A good place to start is below the Scotswood bridge (where the A1 crosses the river). There is easy access, with footpaths running along both sides of the river, you can even park by the river and spot from the comfort of your car.

Norfolk Broads

Classically good otter country but not as easy as some places. Take binoculars and head out at dawn – be prepared to wait.

Bosherston Lakes, Pembrokeshire, Wales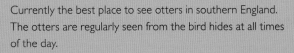

These stunning lily ponds have always been a great location for spotting otters. Go at dawn and dusk. The otters will be about then and the visitors won't.

Bath & Other Cities

Many of Britain's cities now have otters living in them. They are usually not too bothered by people and can often be easily seen, even at night due to street lighting. Behind the bus station in Bath is a good spot!

Shapwick Heath, Somerset Levels

Currently the best place to see otters in southern England. The otters are regularly seen from the bird hides at all times of the day.

Beam Weir, River Torridge, Devon

Sightings are not as certain as other spots but are historically pretty good here. The bridge above the weir is the perfect place to spot from. Dawn is better than dusk.

Philippa: Anyone who cares at all for the natural world will have heard the buzz, or lack of it. I am aware that there is a bee crisis, it doesn't take long to realise the ramifications.

Bees are major pollinators – if they decline our crops and plants don't get pollinated, we don't get enough to eat, the whole understorey of every ecosystem that is the plant world goes into decline and consequently so does the natural world as we know it.

That, of course, will affect our ecosystem too; the river.

The brevity of these facts belies their significance.

But what are we to do?

Well, I have taken a few moments to learn a bit more.

Key facts

- There are many different kinds of bee. They include the honey bee which produces honey to see the colony through the winter. Over the centuries this process has been managed by bee-keepers to provide humans with honey too. Other types of bee, like the bumblebee, do not produce honey. There are even nocturnal bees.
- Bees are the most important pollinating insects, by doing this they enable flowering plants to reproduce.
- Different bee species have different length tongues, which are suited to different flower shapes.
- Flowering plants are dependent on pollinating insects and have evolved alongside bees over thousands of years. Their designs have been constantly adjusted to attract and facilitate the bees' needs.
- The buzzing sound which bees make comes from the way they warm up their bodies in preparation for flight.
- Bumblebees can sting more than once, a honey bee will die if it stings – just out of interest.

How can I help bees?

1. Make a bee home
   - Fun to do with kids
   - good to help populations of wild bumblebees

2. Increase diverse flowering habitat
   - don't use pesticides ✓
   - Plant in clumps ✓
     (need to do more)
   - Use native species with a variety of flowering times ✓
     (Golden Flag Iris, Flood meadow, garden)

3. Keep bees
   - urban beekeeping is an increasingly popular pastime, if people are keeping bees on roofs in cities why don't we have bees here?
   → perfect conditions but what about the river?
   — Call Gareth...

Bees need flowers and flowers need bees.

Philippa: There is a problem frequently found on our rivers that is literally red in tooth and claw. It is the signal crayfish and it threatens our native species, the smaller white-clawed crayfish.

We have had signal crayfish on our river for well over a decade but in recent years have noticed that they have been spreading. You might expect the signal crayfish to come from another river and work its way up the tributaries, but when we first found it, there was no sign of it downriver, only in the highest reaches. Until recently we never understood how this could be the case, but it turns out that the signal crayfish was introduced onto our river by a man who was breeding them commercially. People like this have caused untold damage to the ecosystem which will take years to repair.

We have discovered just how much the signal crayfish has spread.

↓
The kids love searching for crayfish under rocks – but should we let them?

Little by little we have noticed evidence of the signal crayfish spreading. A few years ago we were hunting for minnows with the boys at a section of the river that is narrow and deep and runs through the farmers' fields. Then, a couple of years later, after a heavy flood I was walking the dogs along the river when I spotted a small red mass on a shingle beach bank further down. It was a pile of crushed crayfish shells. I wondered at the time whether it was the flood that had carried the crayfish, or an otter. Either way, it was spreading.

One of the depressing things about making a diary of the river this summer is that we have discovered just how much further the signal crayfish has spread. A few fields upriver from the house are some stepping stones – we are there with the boys, filming some of the plants and searching for fish. I'm standing on one of the larger rocks and looking into the water when I realise that there is a large signal crayfish poking out from underneath.

To eat or not to eat?

For many people the wonderful thing about the signal crayfish is that it represents free food. It is delicious with a little garlic butter. However, the Environment Agency would rather we didn't remove them. One theory is that removing large animals in an inconsistent way just leaves space for smaller ones to breed and grow. But the main worry is that the crayfish plague can be spread unwittingly by moving or handling them.

Only there is not just one – the rock is providing a home for five crayfish.

Of course we pick them up to show the boys. They are fascinating creatures – like mini-lobsters – and the more we pick up the more we see, their large red bodies diving under the rocks. Suddenly I'm no longer inclined to paddle.

Then we remember that we aren't officially allowed to put them back in the river. It is illegal. We can't find anything to keep them in until Richard, our cameraman, brings his son's wellies out of his car. And the alien invaders, with their snapping claws, are placed inside Spiderman boots full of water and taken home to the aquarium for further filming.

If you don't have permission to film or handle them, then legally you should kill them humanely by placing them in a freezer.

The signal may have depressed us by its spread but that is because we understand the significance to our native species. For other river residents it represents an important food source. Not for the first time have we observed otters eating them –

they are almost ever-present in spraint on the river banks.

More significantly, when the river is murky after rain and difficult to fish, we have for the first time seen the kingfisher catch small signal crayfish and even take them into the nest to feed the chicks.

Philippa: It's Friday afternoon, it has been a beautiful summer day, the river is twinkling, distracting, inviting. It is a *Wind in the Willows* day, one that makes the rest of the year pale into insignificance, a day on the river that stays in the memory for ever. A halcyon day.

In the manner of Mole and Ratty we get itchy feet. We don't get days like these very often – blue skies, warm earth, lazy river. So the decision is made for us, a halcyon day is here, we quit work early.

I pack up a picnic – crisps, egg sandwiches, quiche, biscuits – and laden with fishing nets, buckets and baskets of food, we climb over stiles and wander up the river to meet friends.

I spend a lot of time sneezing in the waves of pollen wafting over the fields but otherwise it is a perfect English afternoon. Despite the sun, the plants have paused, as though they too acknowlege that now is not a time for growth but reflection, now is the peak, this quintessential moment is the one we have all been striving for. It is here, we didn't know it was coming but we are all stopping to appreciate it. There is not a sound, the whole world has stopped.

Lately we have spent so much time focused on filming and the wildlife that for this afternoon we need to just enjoy the river as humans. That nearly always involves getting wet!

We stop at a gentle bend, a place where the river slows. Here it is deeper and has lots of gravelly shallows. The bank is gentle and grassy and bathed in sunshine. We gratefully spread out the rug and release laden shoulders.

Someone has to stay with the picnic otherwise Dave, the smelly dog, will eat it, so I elect to stay dry. The boys are in their cossies and into the water in no time. It is colder than they thought, a shock to the system, but the river is irresistible.

They splash each other in the shallows and bravely jump into dark depths (in reality, not even hip height on us). Shivering, they emerge to eat. They grab fishing nets and spend a happy hour just pootling, with nets and buckets, throwing stones, catching minnows and paddling, while we sit and gossip.

As the sun disappears behind the trees, and Arthur starts to get cold, Dave's vigilance pays off – he seizes the remains of the quiche.

For just a few perfect sunlit summer hours the river steals into our hearts and life is perfect.

Test your local river water

http://www.environment-agency.gov.uk/homeandleisure/37811.aspx

Charlie: Timelapse photography is a common technique in wildlife film making – we've all seen those shots of the clouds scudding unnaturally fast over landscapes, or the sun setting over a mountain range in ten seconds rather than an hour – timelapse!

The technique is very simple – A normal video camera takes between twenty-four and thirty frames per second (twenty-five is standard in Britain). These 'frames' are essentially just individual pictures that are all played together quickly in sequence to form a moving image. Timelapse photography does the same thing – it plays a load of individual pictures together in a sequence to form a moving image. The difference is that when you shoot a timelapse you don't take twenty-five pictures a second, you take maybe one every ten seconds or every half hour, hour or day. So when you play them back the world speeds up.

Years ago we used to use fancy cameras with fancy controllers to shoot timelapses but now we just use off-the-shelf digital stills cameras, as the results you get from even fairly cheap cameras can produce high definition images. It might seem odd to shoot a moving image with the camera you normally take snaps of the kids with but it is simple and it works.

Many cameras now come with a built-in interval timer. This is accessed through the menu system and allows you to take repeated frames with a preset gap between each one – that's a timelapse. Other cameras don't have this feature built-in but do have it with the bundled software, so you can plug your camera into your computer and get your computer to tell it what to do. Some camera manufacturers make dedicated timelapse controllers, like the Canon TC80N3 which is fairly cheap and very simple to use. Or you can buy third-party units like the Mumford Time Machine which will hook up to just about any camera and tell it what to do. You can even get timelapse applications for phones these days.

Anyway, once you're armed with some way of taking pictures at preset intervals you can get on with shooting some timelapses. Do some maths first though. The average length of a shot on television is around eight seconds. The average length of any

Timelapse is a very simple and easy way to speed up time.

timelapse I do is therefore eight seconds, give or take a second. To create eight seconds of moving image you need to create two hundred frames (play two-hundred shots in sequence at twenty-five frames per second and it lasts eight seconds on the TV). So you have to make sure you have enough memory in your camera to shoot at least two hundred images, or more if you want the shot to last longer.

Then get a tripod! This is essential. If your camera experiences even the slightest movement while it's snapping away it will ruin the timelapse. So put the camera on a tripod and really 'murder it up' as the film engineer at the BBC used to say (he meant tighten really well so it doesn't move). Next thing to do is point it at whatever it is that you want to timelapse – let's take a cloudy sky as an example. Frame up and then set all of your camera's controls to their manual settings. By this I mean manual focus, shutter, aperture, ISO and white balance. If any of these settings are allowed to think for themselves, they'll ruin the timelapse. Make various test exposures until you are happy with the shot and then set the interval. For clouds in a wide-angle shot, moving at an average speed, I would suggest taking perhaps one frame every four seconds (so set your interval to four seconds). This should give a fairly smooth flow as the clouds move through the shot. If the clouds are already moving fast you could try one frame every one or two seconds. This will speed them up but not too much. The best thing to do is play and experiment. A few rough guidelines appear in the table above.

When you've shot your timelapse you obviously need to watch it. There are many ways of doing this. Some advanced, using professional editing software such as AVID or Final Cut Pro, or others which are cheaper and easier such as Quicktime. There is plenty of software out there so take a look online for one that suits you.

Intervals between frames

Clouds 1–4 seconds
Sunrise 4 seconds
Sunset 2–3 seconds
Stars 20–30 second exposure with 2 second interval (so set timer to exposure time + 2 seconds)
Dragonflies hatching 3 seconds
Sheep walking through a field 1 second
Tide rising 2 minutes

The results you get from even fairly cheap cameras can produce high definition images.

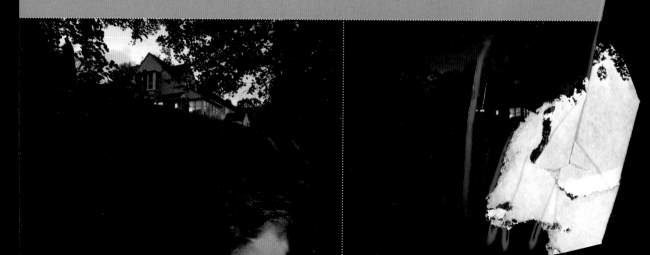

Philippa: I'm sure that every parent is the same, well I hope so, but there are times when I despair of my boys. The things you assume they will want to do more than anything are often the things they shy away from.

Although we are having this wonderful time on the river discovering wildlife, very often the boys get bored, cold or tired quickly and would rather be inside watching TV or playing at sword-fighting with bulrushes instead of appreciating the beauty of the moment. I am a parent with the best intentions, I want to get my children out into the natural world to witness it with all their senses and have real encounters with real wildlife, and yet so often I am, to be restrained about it, frustrated.

It's true our children don't know how lucky they are but then why should they? They have nothing to compare their lives with, so for now they live in blissful, wilful ignorance. And meanwhile we slide between sloppily proud moments witnessing their reactions to nature, and confused desperation when they are simply not in the mood.

However the moth trapping sleepover has all of them engrossed, plus one – Fred's friend Thomas comes to stay because it is the summer holidays.

I have been talking to Richard Fox of Butterfly Conservation about the significance of the river system to moths and butterflies and we are keen to do a species count in different locations around the river. There are certain species of moth that prefer wetland habitats like our flood meadow and

Moths have declined by a third since 1968.

*t – an
th.*

some that spend part of their lives, usually the larval stage, in the river itself. Many, of course, will be food for the trout.

According to Butterfly Conservation, British moth numbers have declined by a third since 1968. Moths and their larva are bird and fish food and are important pollinators, so their decline affects every ecosystem including the river. Richard has been managing the 'Moths Count' campaign to try to get a clear idea of the moth numbers we have now. We decide to help by having a moth trapping night to see how many species we can count. So Richard, this sunny afternoon, brings three traps.

We place them with great excitement on the patio, by the river and in the

Halcyon River Diaries

flood meadow. The children are with us all the way, well, Gus loses the plot a little but then rejoins us. There are different levels of sophistication when it comes to moth trapping: you can buy or make basic traps or just leave the bathroom light on and the windows open and wait to see what flies in. Richard has brought some with an integral light and so once night has fallen all we have to do is flick the switch.

At around 10 p.m. it is properly dark and the children, barring the youngest, are still up and excited (well, it is the summer holidays). We troop out with torches and see what is happening around the traps. There are moths everywhere, we watch them going into the traps and flying around them and the boys are enthralled.

Next morning we spend a happy hour or so moving from trap to trap inspecting all the moths. I never expected them to be so beautiful, so many colours and sizes, so many wonderful names: the 'hart and dart', the 'scorched carpet', the 'clouded silver'. The children are engrossed, writing down the names, identifying them and spotting others of the same species. 'Common footman!, Yellow tail!' they shout. Fred takes some photos and we watch as the moths take to the air.

Crouching in the long grass or standing beside the river, Richard has something to say about all of them. If only they would pay this much attention at school.

Each moth has a story: the peppered moth is coloured to camouflage against its background and before the Industrial Revolution was a light colour to match

Moths are attracted to light for various evolutionary reasons – or because they're just stupid, if you ask Fred.

The moth trap by the pond turned up some good results but the flood meadow trap won the day.

were predominant. In recent years cleaner air has seen a resurgence in the lighter colour. This theory does have its critics but it had the boys recognising peppered moths from that moment on.

Some of the moths are caught gently in specimen jars so that we can inspect them closely, but always released and every single species we identify is written down in the notebook.

The final trap is in the flood meadow and it is here that we find one of the largest and most glamorous, perched against the side of the trap. Fred gently picks it out and we all get the chance to gaze. The elephant hawk moth.

When we have finished gawping at this glorious moth we make a count, and the result is 55 species in one night, 'and that wasn't even a very good night weather-wise because we had some rain so it might have been even higher,' says Richard.

This isn't particularly because we are in the middle of the countryside. We could have been moth trapping in the middle of London without a garden and still have found lots of species. And this is important stuff, the more information and species

the lichen on the trees. But it is thought that when the lichen died out because of widespread pollution and the trees became covered in soot, the darker coloured moths were better placed to survive and the lighter coloured moths, being less well camouflaged, were picked off by birds. Natural selection therefore meant that the dark coloured moths

counts given to Butterfly Conservation the better; they need as much as they can to get a clear idea of what they need to do to reverse the decline of our precious moth species.

Later when Richard has packed up his traps and left, and while I make lunch, the children are really quiet. Any parent will know that this is not a good sign, but when I go to find out what they are doing I discover them crowded round a cardboard box on the landing full of paper moths of all different colours and shapes – they are playing moth trapping.

As I go back to the kitchen I feel good, it's not just the kids who were captivated, I was absorbed too. I simply had no idea that moths existed in such glorious variety – moth trapping will certainly be a family activity from now on.

We could have been moth trapping in the middle of London without a garden and still have found lots of species.

Peppered moths have been becoming gradually lighter again since the dark days of the industrial revolution.

Charlie: Around two-thirds of kingfishers die within a few weeks of fledging the nest. This is due to predation, starvation and drowning. Being a kingfisher is not easy and most birds just don't make the grade.

Andy was one such bird. He was brought to me in a rat trap by my friend Ian. Andy had been caught by (not surprisingly) a man called Andy. Andy had watched the young kingfisher trying to dive into the water, before it started drowning. He had waded into the river and grabbed the poor bird before putting it in the trap (just a suitable cage!)

and giving Ian a call. Ian, who works with me filming, then brought it over.

When Andy (the kingfisher) arrived he was looking bedraggled and one of his wings was hanging in the wrong position. The first thing we did was get him into a cage and feed him. Luckily we had an old chick run for our chickens, a wire box about three feet square. Into this we put

↓
Andy was a few weeks out of the nest when he came to stay.

a few sticks for him to perch on, a baking tray filled with a small amount of water and some minnows. He seemed to take to it well and we filmed him remotely to make sure he was eating enough. Being a kingfisher he was very shy and would get scared if we went near him.

He did well for the first week which surprised me. I assumed that he would die. Keeping birds alive is not easy, they are very susceptible to shock, which will often kill them quickly. Andy seemed to be a tough little cookie. I was concerned about him, though, he was still holding his wing in an odd position and it didn't seem to be getting any better. I contacted an old mate of mine called Lloyd Buck. Lloyd is a professional bird trainer whom I had worked with in years gone by. What Lloyd doesn't know about keeping birds isn't worth knowing. He suggested that the wing needed to be looked at by a specialist so he put me in contact with a veterinary practice in Swindon called Great Western Referrals. GWR are one of the few avian veterinary specialists in the world and the nearest people with the expertise that Andy required.

The next morning I got Andy out of his cage and bundled him into a small cardboard box with a towel inside it. I put him in the car and drove him to Swindon. When we arrived I explained Andy's situation and history to the vet, a young French guy called Minh. Minh said that Andy needed a radiograph (a type of x-ray) and in order to do this he would have to put him under a general anaesthetic. General anaesthetics scare me. I once had an otter who died of heart failure whilst under general anaesthetic and I had visions of the same thing happening again.

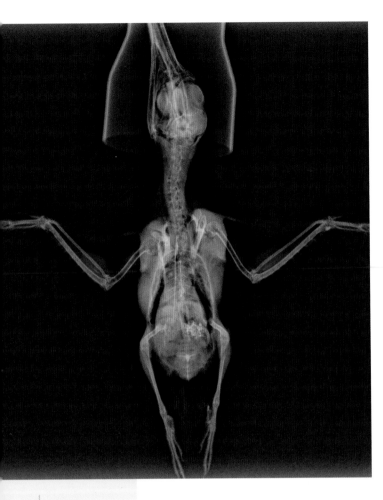

The radiograph shows two breaks to the bones on Andy's left wing joint.

down the corridor into the radiograph room. I wasn't allowed down that end of the corridor due to the radiation coming off the equipment so I stood nervously waiting.

After a few minutes Andy was rushed back in and given an injection to bring him round. He was then gently placed back in his box and left to wake up. The anaesthetist appeared a moment later with the x-rays. He put them up on the light box and Minh examined them. 'It's not good,' he said. Minh could tell from the radiograph that Andy had two quite bad breaks in the bones of his left wing. 'Is there anything you can do to fix him?' I asked. Minh shook his head. 'No, it is not possible to repair this kind of break.' Minh explained the complexities of it which concerned me even more. But he was optimistic, which confused me. 'This kind of injury, if left alone, has a good chance of healing on its own,' he explained. Minh reckoned that if it was going to heal it would take between thirty and fifty days.

The key thing was to try and reduce Andy's use of the wing so we decided it would be best to keep him in a storage box with opaque sides. This would allow light in but he wouldn't be able to see out of it properly which would stop him spending all his time trying to escape. It would also be small so that he wouldn't have enough room to attempt to fly.

When I got home I found such a box, lined the bottom with newspaper, put a nice small log inside for him to perch on and then gently placed him inside. I then gave him a small dish with some fish in it. He took to his new home straightaway and almost immediately cleaned the fish out of the dish. My main concern about Andy's new home was where to put it. Being a plastic box with a lid meant that it could get hot in there so I had to find somewhere cool. I also had a film crew and three kids to keep him away from – too many people

We took Andy upstairs to the operating room. Minh wrapped his hand in a towel and gently got Andy out of the box for an examination. His first thought was that there was a break in the wing. Andy was not happy about being examined and he tried to peck anything that went near his beak. Minh and an anaesthetist then gently put Andy's beak and head inside a rubber cup that was attached to a tube. The anaesthetist started pumping gas into the cup through the pipe. Minh packed cotton wool around Andy's head to make a seal around the cup and very soon Andy's body went limp as he fell asleep. Minh and the anaesthetist then moved very quickly and almost ran with Andy

buzzing about would stress him out too much. In the end I settled on the spare room bathroom, a place few people ever go! The spare room bathroom had been home to an orphaned otter in the past and so it made perfect sense.

Andy did well over the next few weeks. Lloyd had told me to try and keep a regular routine because birds like routine. This was three feeds a day – breakfast, lunch and supper and a clean-out every two days. Cleaning Andy out meant grabbing him first, which not only stressed him but didn't do his wing much good either, so I had to get a balance between cleanliness and recovery – I reckoned two days was a good compromise.

A month later Andy was no different. He was healthy and energetic but his wing was still useless to him. I took him back to Minh who re-examined him. He had a new radiograph machine which gave stunningly sharp pictures of Andy's skeleton and we could very clearly see

would give Andy one more week before taking him back to see Minh – his wing had made no improvement. However, one morning when I went to feed him he was looking rather weak. I fed him and he ate his fish. When I went to give him his lunchtime feed he ignored it and by dinner time he had a pile of fish building up. I became very worried. When birds go off their food it usually means they are in trouble. I watched Andy during the night and he slept on the edge of his fish bowl. He didn't look good though.

The next morning I took him some fish but he wasn't hungry. I picked him up and rather than pecking me he just lay limply in my hand. I tried to give him a fish but he rejected it. I put him back in his box and left him for a while. When I went back to check him half an hour later he was dead.

I was much sadder than I thought I would be about Andy's death. If he had come to me and died a couple of days

Around two-thirds of kingfishers die within a few weeks of fledging the nest.

the breaks in the bones. Minh told me to persist and let time heal the wound. He also told me to cut Andy's feeds down a little as he was getting a bit fat!

I persisted with Andy for a few more weeks. Philippa and I were unhappy about keeping him in a small box, it must have been very boring for him; but it was the advice we had to take if we were going to get him better. As the days went on Andy began to accept me more and would peck me rather than try to escape from me when I fed him – I found being so disliked by him rather endearing!

After three months I decided that I

later I wouldn't have been so upset but, as it was, I had put him through three months of solitary confinement and trips to the vet and all for nothing. By rights he should have died when he was caught drowning in the river. He was a 'genepool loser' as one friend had described him to me. I knew this but I had attempted to beat the system and keep him alive. Staring at him dead in his box, I wished I hadn't.

I noticed something about Andy as I took him from his box for the last time. His bottom beak was just starting to turn red; which meant only one thing – Andy was a lady!

Philippa: It has been one of those evenings. Sometimes managing three boys is like herding cats, and the chaos peaks just after tea, somewhere around bath and bed time.

Tonight, like many other nights, I am craving a glass of wine. I am finally buttoning up clean pyjamas, the milk is drunk, the splashed bathwater is drying on the bathroom floor and teeth are brushed. Scrubbed angels are just climbing into bed with their books and life is just about to get a lot more peaceful.

'Boys,' bellows their father from downstairs, he must have just arrived home, I could kill him. 'Come and see this.'

They need no further encouragement to go back downstairs. The herd of reluctant cats turns into a stampede of wildebeest.

I can hear whistling, happy whistling, surely not my husband, perhaps he is having an affair, he doesn't normally whistle.

As I get to the bottom of the stairs he is looking decidedly smug.

'Evening, my darling.'

Definitely an affair.

'Have a look at what I've got.'

Perhaps not.

He proudly places his rather small video camera on the table in the kitchen. I am intrigued, he is grinning.

'Gather round.'

We all gather round and he presses Play.

There, on the small screen, is a rather familiar empty, muddy ditch, empty that is, except for an apple, but not for long. First some tiny ripples in the water and then a rather fat, dark glossy and very cute water vole appears. A real water vole, not just a sign of one, not just a memory of one but a moving, living, water vole, much bigger than I had thought and recorded just half an hour before in the ditch merrily munching on an apple as if it had not a care in the world, certainly not as if it was one of the fastest-declining mammals in the country.

This is just perfect to show just what is so important about a little ditch but also what is missing from our river. I am thrilled and I'm pretty sure my eyes are shiny as I look up. The kids like him too:

'He's great.'

'Isn't he cute?'

'Come on,' shouts Charlie and in a moment they are fighting with light-sabres in the living room again.

I stay behind for another look. From a TV producer's point of view, our hope of making people more aware of the plight of the water vole just took a massive leap forward. And with my conservationist hat on, that could only be a good thing.

Filming water voles was not easy but did eventually start to pay off.

Philippa: Our hopes for a dry day are diminishing, I can no longer count the number of wet days in a row on my fingers but have had to move on to my toes as well. This is not good. At first it felt like a bit of fun to stay in, a novelty to snuggle up in front of a film with the kids after all this outside time. Can't do the gardening or film in this weather so let's nestle instead. But now the novelty is wearing off, it feels like a winter's afternoon and all it has done today is rain.

I watch the circles on the water outside the kitchen spreading and blurring as they get bigger, blurring just like my filming schedule. The plants I need to film this month are drooping, heavy with raindrops. The lawn is spongy, the willow which was so grateful for the first few days of rain is now officially weeping. The delicate purple petals of the hebe outside my office door lie on the ground, colours fading like spent confetti at a rushed wedding.

What on earth has happened to our summer?

Philippa: The honey in the hive is one of the most beautiful things I have ever seen, luscious in its ooze, true gold. Compared to it, the other kind used to make jewellery is for fools. This gold inspires a movement in the heart, prompts fingers bound in a ridiculous, protective rubber glove to reach forward to scoop it out and taste it.

↓
Bumblebees are not honey-makers but are still very important to flowers as pollinators.

Gareth Baker our rotund bee-man is brave enough to do it. We have known Gareth for years, when he is wearing his bee suit Charlie describes him as a Teletubby. He can talk about bees until long after the sun sets – he is passionate and fearless. I can only look with envy from inside my bee suit, too paralysed by fear of being stung to remove my glove, as one of his big digits scoops a large dollop and places it on his tongue. I am so tempted but furious buzzing prevents me from baring even an inch of my flesh.

I had been thrilled when we decided to get bees as I have always fondly remembered staying with my beekeeping aunt and uncle as a child. Their bees were a mystery, kept in the orchard behind a

hedge at the far end of their garden. I was always advised not to go there, advice which I heeded, but I would peer through the gate and wonder what was going on in those hives. At harvest time, though, the mystery was revealed. I loved to sit in the kitchen when it was time to make the honey. I watched with the same curiosity that I watched the vicar and his acolytes prepare for communion, but I was a lot less bored. It was so ritualistic – the clothing, the special equipment, not blessed but sterilised none the less and treated with a certain respect. The kitchen would be transformed into a human hive; the annual and familiar process utterly absorbing my aunt and uncle so they barely spoke as they worked. And the sublime result, jars and jars of golden liquid.

Bees are essential to any habitat – their work as pollinators is something that we all ultimately rely on. Without bees we have no food security, no sustainability.

Yet the bee is certainly not the first species that comes to mind when you think of the river: mayfly, perhaps, dragonflies but not a bee surely …

Gareth believes that the gold dripping onto his tongue is the best kind; not pure but a mixture, a mixture of different sources of nectar.

He has kept bees for many years and now runs his own company, Wye Valley Apiaries, which looks after hundreds of hives 'from Monmouth and the Forest of Dean through the farms and orchards of Herefordshire and beyond.' He has hives in cider orchards by the Wye, in heather up

Without bees we have no food security.

the mountains and now here, in our back garden because he has agreed to help us by seeing how bees benefit from the river.

The nature of the countryside now means that often his bees are limited in their options, in terms of plants to visit. These days farms go for large expanses of one crop, acres and acres of monoculture in many places having replaced acres of diverse planting. So the bees have masses of one crop, for example oil seed rape, for one period and then have to move on to something else. Now this doesn't make bad honey but Gareth believes that the best honey comes when the bees can select nectar from a variety of flowers and they will go to a surprising amount of different flowering species.

This first dawned on me when Gareth came with the hives and we started to search for the best place to put them. We had wandered into the vegetable garden as we were looking, 'Well, there isn't much here for them,' I said. 'I expect they would prefer the flowers in the other part of the garden.' But Gareth pointed out the small white flowers on the broad beans, 'Oh no there's loads – look there, bees love those,' he said in his soft voice.

We wandered around the whole garden and I assumed he would choose the most flowery part, the nicest part of the garden with wonderful beds of flowers but in the end he chose to site the hives in the wild meadow.

'It's away from the children and any disturbance and surrounded by flowering

Gathering nectar is thirsty work and the bees would often stop to drink from the pond.

plants.' Sure enough, it was – they just weren't as showy as the ones in the flower bed. 'It's also close to the river and all that bankside vegetation.'

As I looked at the line of trees and bushes stretching over the fields of crops indicating so clearly the path of the river, for the first time I saw it differently. If it weren't for the river all that vegetation would have been cleared for crops, the river was to be a kind of high street for bees, they could buzz up and down perusing and when they found something good on offer, return to the hives and tell their friends about it.

Sure enough, just a couple of hours after we had installed the hives and waved goodbye to Gareth's Land Rover, I glanced up from my weeding in the vegetable garden to see that the new honey bees had already ventured over 200 yards and were working the broad bean flowers.

Since then we have been spotting bees all over the place, drinking at the weir, and, as we expected, visiting the flowers in the garden's borders but also up and down the river. Just as Gareth had

predicted, the bees use the river as a type of motorway where every plant is a service station. River banks are places that still hold a wide variety of plants all flowering at different times, from the smallest wildflower hiding under the trees, to the golden flag iris in the sunshine with its feet in the water – and the trees themselves, a mixed bag all providing pollen and nectar to the bee. And if Gareth's opinion was anything to go by, providing them with the wide variety of pollen types they need if they are to make the best kind of honey.

And a couple of months later here we are standing in the flood meadow, both now kitted out in Teletubby outfits and gloves, buzzing bees, wood smoker and one hive open. And there is Gareth's big finger dipping into the golden ooze, making me jealous that I haven't got his bravery. The thought of exposing my delicate pink skin to those bees after prising apart their house to peer inside at their private lives was too much.

They have done well, these bees in the first hive, it wasn't full yet but nearly.

We put it back together and move to the next. Gareth removes the top boxes, I have an itchy nose under my veil but ignore it and peer over the edge of the hive. After the glorious plumpness of the racks in the last hive this is a disappointment; rack after rack of unfilled cells, nothing to see. Gareth goes quiet. He never goes quiet, his soft voice is a constant sound like the soft buzzing of the bees, he must be concerned. He immediately removes the boxes to take a look at the bottom one, the breeding box. One by one Gareth removes each rack, peering at it for a long time before replacing it. The initial shock over, he has begun to talk again.

What I'm hoping to see is a queen or any evidence of one,' he says slowly. He is concentrating hard on the cells, checking each one, 'I'm looking for young larvae that have hatched into the cell but I don't know if you will be able to see them, they look just like someone has snipped off the smallest length of the finest white cotton and dropped it into the bottom.'

I peer too but my nose itches, my hair falls in front of my eyes and the veil obscures everything.

We find no queen and no evidence of one, she must have abandoned the hive soon after they arrived. And when we open up the next hive there is very little honey there too. Upon further inspection, they do have a queen and are concentrating on breeding, rather than honey-making so at least they do have larvae.

We move a rack of these into the second hive. 'They will take them on from there,' says Gareth, 'and hopefully rear a new queen.'

We leave the hives on a bit of a low, perhaps the magic of the river doesn't really extend to bees after all. I can't help but wonder what made the queen leave.

Was our riverside meadow simply not good enough?

But what am I thinking? It isn't, of course, as simple as that. The sophisticated hive politics that makes grown men like Gareth, who had challenging jobs in the city, spend a lifetime peering into hives, on a more significant and absorbing voyage of discovery. Swarming is a natural part of bee behaviour, a way of reproducing and increasing their numbers but not good news for the beekeeper because the vast majority of the productive bees are lost. It is something therefore to be avoided.

There are many reasons why a new queen did not emerge: treason, murder, predation – who knows what has been unfolding down in the wooden boxes in the flood meadow? Only two things are sure: we will have to wait a lot longer for a honey harvest, and I am still scared of being stung.

There is still time for the hives to recover, however, and the Himalayan balsam is only just beginning to bloom.

Philippa: **Our only native crayfish is in decline across Europe. The white-clawed crayfish can live up to fifteen years in streams, rivers, canals and lakes, hiding amongst stones or tree roots. They are important to our river ecosystem because they clean up the river bed, eating everything from dead fish to plants and detritus.**

Like so many other endangered species, the problems really start when there is more than one threat. As well as facing competition from the signal crayfish, the white-clawed species is also threatened by disease: the crayfish plague. As a consequence, the white-clawed crayfish is now protected and is a priority species under the UK Biodiversity Action Plan.

Crayfish plague is a fungus-borne disease from North America. It is virulent and causes death within a few weeks. It is introduced to the water by signal crayfish and can be carried by fish or even equipment or boots that have been in contact with animals. It can kill all the white-clawed crayfish in a river within weeks.

The Environment Agency has set up ark sites – places far enough away from rivers infected with crayfish plague to be considered safe. It is vital that any equipment which has come into contact with a river where there are signal crayfish is thoroughly cleaned to prevent infection.

As part of our diary, the boys and I are privileged enough to be taken to an ark site to see the white-clawed species for ourselves. We scrub our wellies and buckets before arriving at the secret location and once we find the small crayfish, take some photos. I wonder whether we will have lost the last of the white-clawed species by the time the boys are grown up. Hopefully by then we will have found a way to eradicate the plague and remove the signal crayfish.

When we return home my optimism kicks in and with fingers firmly crossed we begin to search the further reaches of the river. Sadly, although we look and look, we find none left on our river or any of the tributaries. Just many hundreds of signal crayfish.

White-clawed crayfish are smaller and more delicate than their American cousins.

White-clawed crayfish
(*Austropotamobius pallipes*)

Philippa: A rainy August evening, a small group of us in a church hall gathered to play with polystyrene sheets and clay.

↓
Mink rafts are a very simple but efficient means of monitoring mink.

Fred and Gus spin around and around us, so full of energy even though it is way beyond their normal bedtime. They are in that wonderful place where the summer holidays mean any concept of normal time comes to an end. One long day rolls into another. School or any responsibility is just a strange end point far off in the distance, so far that it isn't worth thinking about.

Robin, who helped me find the water voles in the first place, has gamely offered to help further and has organised a whole evening to get us moving on to the next phase. We are here to build mink rafts.

The threat from the mink needs to be monitored and, if necessary, action taken. The action is the worst bit of this project

and that is all I really want to say about that at the moment, more later.

As part of the natural ecosystem, water voles have defence strategies against predators – tunnels. When an otter, heron or other predator appears the water vole takes refuge in its tunnel system. The entrance is underwater which deters most predators. The otter, who is capable of the dive, is far too big to fit into the tunnel.

The mink, however, can both dive and get into the tunnel and once in there can wreak havoc, potentially destroying a whole colony of water voles in a night. The water vole has had thousands of years to evolve its defence strategies against the predators which belong on the river but no time at all to begin to find a way to evade the voracious mink who only arrived a few decades ago, not even a blip on the evolutionary scale.

If we are to protect the small colony of voles on our river, and even increase their numbers, the mink must be controlled. A system is in place in the West Country which is working.

The British Association for Shooting and Conservation (BASC) is full of active volunteers who are 'passionate about the countryside and wildlife'. Yes, they shoot and fish but they kill food to eat it, not to waste it. It is in their company that I find myself tonight, for these are the volunteers who have turned up to give us, well actually the water voles, their time. I am struck by this donation; time is the most precious thing I have. I find it extraordinary that these men and one lady care so much about the water vole that they will come and sort the mink out, and this is a project that is neither simple nor for the faint-hearted. This is how it goes.

Firstly we build some 'mink rafts' (tonight here in the village hall). These are polystyrene and corrugated plastic rafts bolted together with cable ties. In the middle is sunk a basket containing florists' floral foam which stays damp when floated in the river and consequently maintains the moisture in the clay pad on top of it. Over the top of the clay pad goes a removable tunnel (made of corrugated plastic).

The mink raft has two modes: as described above, it is in monitoring mode. It is tethered to a tree and floated in the middle of a river where it relies upon the natural curiosity of the mink. The clay pad is regularly checked by one of the volunteers and if and when mink footprints are discovered, the raft is switched from monitoring to trapping

mode as quickly as possible. In this mode a trap replaces the clay pad, a cage which can contain a mink but not hurt it. At this stage the raft needs to be checked at the very least once every twenty-four hours, preferably at dawn because the mink is a nocturnal creature so the likelihood is that it will have been trapped at night. It is cruel to keep any animal alive in a trap for longer than is necessary and so it is critical that the volunteer checks the trap promptly.

When the person monitoring the raft finds a live mink, they must kill it. It is illegal to release it back on the river.

Simple, eh?

There are two difficulties.

Firstly, it takes time and dedication to monitor these rafts regularly. This is critical because one mink can cause devastation in a water vole population very quickly. There need to be volunteers prepared to do this, and back-up teams for when they aren't available.

Secondly, the mink has to be killed. Volunteers accept that this is necessary,

and attempts to recreate the scene in *Ghost* – consequently he doesn't concentrate and is the last to finish his raft which is considerably scruffier than mine.

But all along the gruesome thought of what these rafts are for is in the back of my mind. I know what animal lovers across the country will think – 'murderers'. I will think it too. But what if we did nothing and let our native species of water vole die out? Surely that is more neglectful.

One mink can cause devastation in a water vole population very quickly.

and will do it. I also accept the need to do it but sincerely doubt whether it is something I could do. Bloody hypocrites: the world is full of them, but this one is me.

We spend a jolly evening creating our mink rafts and shouting at the kids. Charlie jokes around a lot with the clay

I glance at Robin, who is busy showing Charlie how to build the shelter over the clay with corrugated plastic and cable ties, his eyes sparkling above his beard as he laughs. His wife Pam is elbow-deep in clay and sand, also smiling as she explains why one volunteer has clay the wrong consistency of chocolate mousse.

Yes, they are hunting and shooting types, and these are people I have often frowned upon, but I realise neither of them stride around the countryside shooting things for fun. Robin has spent a lot of time gathering volunteers and setting up this evening, bringing everything that we need and co-ordinating other similar groups across the West Country. He does this for pleasure, the pleasure in knowing that he is making a difference to the countryside that he cares so much about. For him shooting the mink is about balance, it is something which has to be done.

It's all very well for me to be squeamish, but am I being precious? Surely I can't stamp my feet in anguish at the plight of the water vole if I am not prepared to pull the trigger?

But when I think of killing an animal it goes so against the grain that my brain starts churning over again. I simply hate the fact that we need to kill mink.

Humans, of course, are the cause of the mink's problems. If we hadn't brought it from North America it wouldn't be causing havoc in our delicately-balanced river ecosystem. What we are having to do now is reverse a mistake made decades ago.

Alien species have caused unpredicted problems in delicate ecosystems all over the world. In the Galapagos islands feral goats originally released by whalers consumed brush, removing the habitat of finches and the food of tortoises and exposing the land to erosion. After eight years of shooting them the goat population on Santiago has been eradicated. It has been considered a conservation milestone and the only way to recover the ecosystem on the island. Needless to say, the shooting was done by trained hunters as humanely as possible but the goats had to be killed nonetheless.

Goats, rats, cats, grey squirrels – whenever an alien species is introduced there is no kind or easy way to stop the problems they create for native species – if there were it would have been taken. There is no other option.

It is an uncomfortable reality, one that I would rather not think about but if we lose the water vole, then it is lost for ever, and it is currently our most endangered mammal. We must take responsibility and we must act fast. When it comes down to it, however uncomfortable I feel, I cannot stand by and watch us lose a species because of our stupid mistakes.

At the end of a successful evening of building, Pam and Robin talk us through what we might expect to find in terms of tracks in the clay and they show us some examples. Of course, there won't just be mink tracks – every curious

creature will leave a trace of their visit, from otter to moorhen, perhaps even the odd water vole!

We gather round a large map of the area to work out the best places for the rafts and allocate plots to volunteers.

Outside the church hall the cloud has lifted a little and the rain has stopped. Ready for bed we bid farewell to our new friends. Needless to say, although we are shattered the boys are still running around like spaniels so we head to the pub for lemonade and crisps or something a little stronger.

↓
Not all the hives have been a success but at least one drips with honey.

Philippa: **It doesn't take long to nick all their honey. Well, that is a bit what it feels like.**

Gareth and I go through the ritual that is now becoming familiar, don the Teletubby outfit, nose starts to itch. I watch as Gareth ravels lots of string and bits and assembles them with care in the body of the smoker. He adds a pine cone, lights it and blows it to embers and the funnel- shaped lid is replaced while we all enjoy the smell. The smoker is symbolic of the beekeeper and the smell will now always be with me and remind me of our bees.

And so we visit the hives on this September morning with purpose. The bees are calm and quiet because although the sun is out it has been cold and I am excited because I am finally going to get my paws on some of that honey.

There is much anticipation as Gareth levers off the lids but inside little has changed. Hive one oozes; hive two now has encouraging signs of reproduction,

house is quite different from my memories of my industrious aunt and uncle, calm and relaxed. We have a film crew cluttering up the place – Richard and Steven stick their fingers into the mixture at every stage – and a toddler and occasional husband, as well as Gareth and myself. The atmosphere is noisy and joyous, I wonder if it is a sugar high or just the gold rush.

First we set up the equipment: an extractor, and a heated filter unit all on my old butcher's block. The extractor is a plastic barrel with apparatus in it which holds the frames and spins them so that the centrifugal force removes the honey. And the heated filter unit is a large stainless steel tray at an angle with hot water underneath, kept hot by a kettle element. At the top is a piece of wood with a hole into which the frame is designed to wedge while we cap it.

The smoker is symbolic of the beekeeper and the smell will now always be with me.

there are new larvae and a queen, but only a smidgen of honey not worth nicking. And hive three has begun to make some honey but not a huge amount. Each hive has its own character, is its own world. Still, no matter, the end of the summer is here and it is time to harvest, what they make from now on they can keep to see them through the winter.

Gareth carts the honey frames into the house.

What fun a few hours a kitchen full of honey can be. Honey making in our

We remove the first frame and we all taste the honey now, the longed for scoop of the finger through the honeycomb, sweeping the wax caps off and freeing the liquid honey. The sweet taste hits the tongue and then rolls right around the back of the mouth, before it begins to melt and other flavours are captured by the tongue. It causes us to grin at each other like idiots and forces us to take another scoop. Gareth watches us, calm in the manner of my aunt and uncle, it is of no surprise to him, he has seen it all before.

Health benefits of honey

- Antibacterial
- Antimicrobial
- Easily digested and natural sweetener
- Great for burns
- Low fat
- Contains variety of vitamins and minerals
- Contains antioxidants
- It is a humectant, attracts water so great moisturiser for the skin
- Possibly has a beneficial effect on the good bacteria of the gut

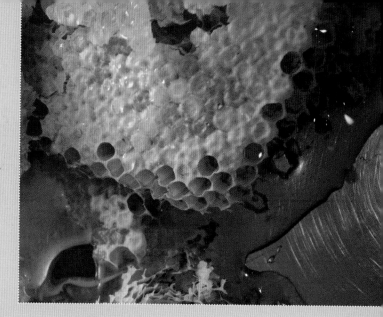

We firstly take the caps from the cells by scraping them on top of the heated filter. The mixture of wax and honey falls, folding down over and over onto the plate and as it is warmed the honey flows further down beyond the filter and into the waiting bucket below. For a while it is intercepted by the tongue of Arthur who is at exactly the right height to catch the honey just after it comes out of the spout.

The rest of the frame goes into the extractor, eight frames in and time to work the arms. Around and around we spin it, taking turns because our arms ache, (you can get mechanised ones) until eventually all the honey is out.

Nothing is added to the honey, nothing is taken out, it is simply removed from the cells and at the end we have a huge barrel of honey with a tap on the bottom. I sit on a stool with a box full of empty jars and fill them. My friend has come to visit, the honey has attracted more people to the kitchen, she helps me jar it up. We smile all the while.

Gareth wouldn't change his job for the world and I can see why.

From the three hives, with two of them not really concentrating, we have produced over fifty jars of honey. It is light in colour and subtle in flavour. Gareth says it is characteristic of polyfloral honey – a honey made from a variety of pollens, particularly balsam.

Gareth wishes we wouldn't take photos until the air bubbles have come out of it, he says that if we knew about honey we would know that we should leave it to settle. But we don't know about honey and we can't resist and so he just stands and talks while we eat and film and photograph the precious gold from the flood meadow.

Each jar is a miracle, this natural, health-giving ingredient which I can use for so many things has been gathered from the flowers up and down the river and in the garden outside and processed inside my home. Each jar is a signature of the summer's flowers. This is not like any other practical conservation work I know. This is going beyond conservation and reaching a place where our relationship with nature is still intact. I am still frightened of being stung but more frightened of what we would all miss without experiences like this.

I go back to my notes and lists.

Number 3: Keep bees, was certainly best on my list of things I could do to help.

Number 4: Keep keeping bees.

Honey cookies

My Auntie May died recently but the imprint of her home cooking, bee-keeping and honey-making will always stay with me. Her many cookery notes were passed down to me along with her preserving pan. I read them with joy.

Alongside the entry forms and rules for the bee-keeping competitions, this was one of the many recipes from her notes. It exploits all the potential of honey for sweetness and nutrition in an easy-to-make, home-made cookie that will grace every snack box with love, if they last that long.

**3 tablespoons honey | 100g margarine | 1 tablespoon water
1 tablespoon soft brown sugar | 180g self-raising flour
100g rolled oats | 100g dessicated coconut | 1 beaten egg**

Set the oven to 150°C/300°F/Gas 3) or middle of baking oven in an Aga.

Melt the honey, margarine and water together in a large saucepan. Remove from the heat and mix in the flour, oats, coconut and sugar.

Mix in the egg.

Roll the mixture into balls the size of a ping-pong ball and place them on a lightly greased baking tray, pressing them down with a fork.

Bake for 15 minutes until pale and golden.

Philippa: A key question for naturalists when they look at an ecosystem is: how does it support the top predators? As we watch our otters and kingfishers fishing we often wonder just how much food is there for the taking. Today, thanks to some intense research, some very sophisticated technology and the help of a top government agency we've come a lot closer to finding out.

One of the major problems of studying a river in such depth (sorry but couldn't resist it) is that you can't see clearly what is happening under the water – so the most productive and important part of the ecosystem is out of sight. We partly get around this by using our big tank, taking great care to make it exactly like the river bank, enabling us to observe close-up fish behaviour and the voracious habits of certain insect nymphs.

We have also placed cameras under the water in the river itself giving us remarkable insights into how an otter hunts, and surprising us with the fact that there are signal crayfish in the meadow part of the river. In the end, though, we are hampered by the lack of water clarity and the fact that neither of these methods enable us to get anywhere near a complete picture of underwater life.

It is fascinating that a whole family of otters, who in captivity need at least a kilo of fish a day, will linger around for weeks, their presence indicating that the river is healthy and has enough fish to support them. But what do they eat exactly? How many fish are there in the river and what sort, and do we have in the river any of the otter's favourite food, the endangered eel? How are we going to find out?

I am not the only one asking these questions. The Environment Agency, the body charged with the care and protection of our rivers, survey every river as regularly as resources allow to gauge their health. When I investigated the results of the last survey on our river about four years ago I found it had not done every well. When I contacted the agency to tell them about our project they were thrilled and offered to help and show us how the professionals study underwater life.

So in late August we went fishing, electrofishing. I took Fred along for the day because, like most boys of his age, he loves catching fish and the prospect of doing so with electricity was too good to miss.

Fred and I are confined to the bank for safety reasons and stand and watch Geoff and his team slowly and methodically set up their machinery.

With the main generating kit in a dingy and one man to guide it, four people enter the water: two with electrodes which they sweep under the water, and two with nets. The noise of the generator fills the air but aside from that it is a typically beautiful summer day, the river smooth as a mirror. Suddenly the surface is littered with bodies, lifeless bodies. It is a disturbing sight, Fred and I hold hands but the team don't flinch and efficiently catch the fish and pop them in a large container of water in the dingy. By the time we edge down the bank to peer in they have revived and are swimming around. They have just been stunned for a few moments.

Now that we feel a little easier about the process, we start to enjoy it; fish of all shapes and sizes – from tiny bullheads to medium trout – bob up. After about twenty minutes the team take a break. They are wearing rubber drysuits, in

The information we gather will be used for years to come.

the heat it is hard going so a break is welcome but also the tank is filling up fast with fish. We measure and record each one before releasing it unharmed back into the water. Fred learns the difference between chub and dace, and a large trout comes in at a whopping 42 cms.

'I really didn't think we'd find anything that big,' says Geoff, 'we certainly didn't last time.'

Threats to the eel

- Commercial fishing
- Climate change leading to changes in oceanic currents
- Pollution
- Barriers on rivers which block migration
- Predation

Good news for the river. But sadly there are no eels.

We search several more locations through the hot day and Fred gets to be a dab hand at holding a flipping fish to the measuring chart before releasing it. But although the amount and variety of fish species we measure is great both above and below the weirs, there are no eels.

So the Environment Agency offers to come out again, this time with a very special piece of kit, a sonar camera. This is high-tech, mainly used in industry to inspect hulls or underwater structures, and the agency only have a few available to them but it is proving invaluable in their studies, particularly of eels in other catchments. Because it is sonar, using acoustic signals which are bounced back, it can still work in murky and dark water where our normal cameras wouldn't see a thing. We are lucky to be able to use it.

The camera doesn't look very impressive when it arrives, an unassuming black box, and it takes half a day to rig it up, off some scaffold poles from the side of our patio. Its field of view is the mill pond and the sluice and it will pick up any movement and record it onto a computer.

As soon as it is running we begin to study the images but, just like an

I struggle to make out the sonar underwater scene on the laptop.

ultrasound scan, it is very difficult to interpret. To give us an idea of scale and shape we persuade Charlie to don his wetsuit and mask and dive in and out of the freezing water in front of the sluice gate. It helps but to be brutally honest it is mainly so that we can have a good laugh at him in his wetsuit.

The rig stays for three days and the laptop faithfully records all that passes. Boring really.

Not long later I go to the Environment Agency laboratory to see the results. It is fascinating. Again it takes a while to get the hang of what the images are but soon wriggling before me I realise I can see large fish swimming back and forth, and then something else on the bottom. I know immediately from the movement what it is, and look at Pete,

'A crayfish.'

He nods, 'Afraid so.'

I hadn't realised they were now in the river outside the house.

But then something huge swims by.

'What is that?' We measure it using the computer and it turns out to be a 70 cm long fish, which we conclude could only be a huge brown trout. Not just fantastic to see but also scientifically significant because it means that the river has spawning trout in it and is enabling them to grow to a good size – lots of otter food.

And then just visible in the depths we spot a small eel hanging out on the very bottom of the river in front of the sluice. Small to my eyes but actually 35 cms long, long enough to be a female waiting to return downriver to the sea and head halfway across the world to spawn, just waiting for a dark, moonless night with a storm to raise the water level just a little. Significantly it is waiting just above our weir. Proof at last that the eels can navigate this far.

And then the best news of all, Pete tells me that when the Environment Agency surveyed the river four years ago they found on average three species of fish in each location. This time that number has jumped to eight, which means the river in terms of its health can be upgraded to 'good'. Great news for the otters and for us.

The very next time it rains all day, and the river roars as it falls over the weir and through the sluice and the clouds cover the moon at night, I look out of the window and think of that female eel and wish her well at the start of her amazing journey across the world – unless, of course, she gets eaten by an otter on the way.

The European eel (*Anguilla anguilla*) – critically endangered

- A life cycle of transformation and travel
- Born somewhere in the Sargasso Sea south of Bermuda. The exact whereabouts remains a mystery to science
- Small leaf-shaped larvae drift to Europe on the Gulf Stream
- Transform into glass eels, so called because they are transparent but now eel-shaped
- Migrate upstream into rivers when the river water is between 10 and 12 degrees, during this journey they get their first colour and are known as elvers
- Spend between seven and twenty years feeding and growing in our rivers although they can remain for longer if the way back to the sea is blocked
- Females grow much bigger than males
- On reaching sexual maturity the eel changes into a silver eel, dark back, white belly and big eyes
- On a moonless cloudy night when the river is slighty higher, often during a storm, they will commence their final journey
- Spend the last six months of their life travelling 6,000 kms over the Atlantic Ocean and to the Sargasso Sea to spawn and then to die
- Eels can live up to eighty-five years although this is rare

Halcyon River Diaries

Autumn

5th September Blackberry picking

Philippa: Lately I have noticed dark blackberries in the roadside hedgerows but because I have been driving I haven't really had time to give them a second glance, it seems a little early to me.

But all the week the thought that there might be plump blackberries waiting, free and bountiful, niggles at me. So on Saturday morning once I have fed the voracious machines of the house (dishwasher and washing machine), I gather up the boys and a couple of pots just in case and we head over the bridge and down the bank to the blackberry bushes that lie between the river and my small orchard.

And sure enough there they are – and not just a few, there are tonnes of them. I have never known so many so early and am filled with excitement, just to think I might have missed them. We raid the bushes, the boys' pots remain empty

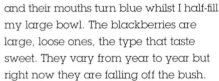
BLACKBERRIES

and their mouths turn blue whilst I half-fill my large bowl. The blackberries are large, loose ones, the type that taste sweet. They vary from year to year but right now they are falling off the bush.

The boys, having sated their appetite, quickly lose interest and, abandoning unused bowls, begin to play a game around the trees. The water in the river is low and so the noise of the weir is more of a chatter in the distance and I pick and pick. Driven by greed I yearn for a pair of stepladders to get those most juicy and most high-dangling, taunting against the blue sky.

I smile as all those familiar blackberry-gathering thoughts come to the fore of my brain: 'If I can just lean in a bit further then I can just ...' I think the same things every year. 'If I get that pile of stingers out of the way then there will be some really big ones lurking all around there' or 'I'll just fill this bowl and then I'll stop' or nasty

↓
*Out of a handful of
blackberries only
a few will make it
to the kitchen.*

Blackberries (*Rubus fructicosus*) are otherwise known as brambles.

more competitive ones like, 'There's no way I'm leaving these for someone else to gather'. Sometimes I wonder at what comes over me, I seem to be driven – the more I gather, the more I want. I lose track of time and all I can see are these nutrient-rich berries that cost a fortune in the shops, waving at me. The thought that turned out to be the most fatal today was, 'I'll just go up the top paddock and see if there are any there …' The boys go to find Dad.

Abandoned by all, completely alone in an overgrown field with nothing but denim shorts and crocs for protection against the nettles, the urge to pick took over, I was done for.

These foraging instincts must stem from some part of me still related to paleo-woman, women must have been slicing open their skin on thorns and battling with stinging nettles to satisfy their instinct to gather 'just one more bowl' since they first appeared on the planet. That is probably why when I give myself over to it, it is a deeply satisfying pastime.

We have certainly been doing it for thousands of years, there are references to blackberries as far back as Viking literature and berries still form an important part of the Scandinavian diet. Fans of the paleo-diet, which follows the principle that if our ancestors didn't eat it then neither should we, would be

Winter blackberries

These blackberry portions will ooze onto winter breakfasts when the cobwebs are sparkling with frost on the bridge and the children have porridge to warm their bellies before grumbling off to school. They are also great for after supper with ice cream or with a crumble topping.

600g blackberries (no spiders) | **1 squeezed lemon** | **4 dessertspoons of honey**
(These quantities can be adjusted depending on how many blackberries you have.)

Put all the ingredients into a saucepan and gently bring to a simmer for ten to fifteen minutes, try not to boil blackberries for too long so that they keep their shape.

Leave to cool and then freeze in portions, flexible muffin moulds are best for this. When they are frozen bag them up.

The best thing about these blackberry portions is not just that they are beautiful when frozen but that they have the power to spin you back to the last days of summer as they touch your taste buds.

I love knowing that the summer goodness continues …

nodding their head in approval as I graze my navel, reaching ever higher. These are the kind of foods our bodies evolved to eat, this is back to basics.

My fingers are covered in thorns and juice, this is a painful process, so why keep it up? Well, our ancestors knew a thing or two, every little berry is a bargain, approximately thirty for the price of one. Because each berry is a collection of drupelets, little berries, they are high in dietary fibre, all that extra seed and pectin adds up.

I shove some brambles out of the way to stand on a thistle, and a blackbird has the gall to tell me off. This protective thicket is probably her home. Once she is quiet again there is just the buzz of insects.

Paleo-woman certainly didn't know about dietary fibre, or the high level of antioxidant compounds which are anti-carcinogenic. But I bet something in these berries' shiny plumpness attracted her.

The sun warms my shoulders as it would have done hers. I lose another blackberry – too ripe to wait for my fingers to close around it, it relinquishes its hold on the bush at a mere tremble of the branch and lands beyond my reach on the ground.

I can hear the kids coming, moving in, a chatty weather front. The noise gets closer.

'Dad says it's time to come in for lunch now.'

'Ow! Ow! Ow! There's stingers.'

'Mum, Arthur's got stung.'

Time to call it a day, four pots overflowing. I am sure of one thing, paleo-woman would have been just as reluctant as me to turn away from those ripe vitamin drops, that only come once a year.

'Have you got berries?'

'Yes. Where did you get stung?'

He can't answer because he has just crammed as many blackberries as he can fit into his mouth. Suddenly the pain is gone, and his hands are blue, I can't help but smile, I know the feeling.

In World War Two children, including evacuees, were called in as a workforce to pick blackberries, which the Women's Institute turned into jam by the tonne. One woman in Northern Ireland apparently made 4,897 lb of jam on two primus stoves in the bedroom of her bungalow. When the pressure was really on to stop relying on imported food, the Ministry of Food issued recipes for jam and advice on what and how to pick and we didn't need to look any further than our own hedgerows for an incredible source of nutrients. Just three handfuls can provide 100 per cent of our daily vitamin C requirement.

I glance down at my arms and legs, I look like I have been in a fight.

I wander back downriver to the kitchen with brimming bowls in my arms. More than just my tastebuds have been satisfied by my annual raid, I have reconnected with the women before me, that reached over, under and into the spiky brambles and suffered this pain for one reason only – all of them knew a good instinct when they felt it.

Blackberries freeze well and smell of summer days as soon as they start to defrost.

Blackberry sorbet

I would never have really bothered with sorbets but for Arthur, he has problems with eczema which is really exacerbated by dairy produce, so I have got into the habit of having sorbet standing by for him when the others have ice cream.

More and more the others have chosen his sorbet over their ice cream and I have found that sorbets are a wonderful way to use up fruit, and are fun to experiment with. This sorbet is just what a sorbet should be: bright and crisp on the tongue, (or round the chops if you are three-year-old Arthur).

300g blackberries | 100g light brown sugar | juice of half a lemon about 100ml of water | 2 sprigs of mint (optional)

Mix the sugar and water and boil until the sugar dissolves.

Puree the blackberries and lemon juice together and push through a sieve to remove the seeds. Mix with the syrup and allow to cool. If you like, add a couple of sprigs of mint while the mixture cools and remove later.

When the mixture is cool pour into the ice-cream maker or into a round plastic bowl in the freezer. A round bowl is best because you need to take it out every fifteen minutes or so to whisk the freezing slush. When it has reached a consistency you like, allow it to freeze completely.

Serve garnished with sprigs of mint or whole blackberries.

Blackberry cobbler

Overlook this at your peril: a simple but robust, country-cooking, body-filling pudding. This won't bloat you out, instead the warmth and taste of it lingers, reminding you for long afterwards of just how good it was.

It is easy to make, and just like crumble, gives you the reward of smiling faces around the table; everyone loves it, particularly the children and everyone thanks you with clean bowls.

120g melted butter | 165g sugar | 150ml milk | 2 teaspoons baking powder
125g flour | 400g blackberries | 2 tablespoons of honey

Preheat the oven to 170°C/325°F/Gas 4 and melt the butter.

Mix the sugar, baking powder and flour together and then add the milk. Combine two tablespoons of honey with the melted butter and add to the mix. Stir to make a smooth batter.

Place the blackberries into the baking dish and then pour over the batter.

Bake for one hour until golden, serve hot with rich vanilla ice cream.

One of the best things about this dish is that leftovers are really good cold the next day.

Philippa: I received an e-mail from Robin Marshall Ball today.

Inbox

Hi all,

Chris Holloway has photographed tracks on his raft … it seems that the water vole colony is extending upstream from their original ditch. The tracks are of an adult vole.

Paul, very remiss of me but when we visited your raft a few weeks ago there were water shrew tracks on it … very small and indistinct but in the middle of the stream, it couldn't really have been anything else.

Many thanks for all your efforts … this can only be a 'good news story'! Keep up the good work!

Robin & Pam

This is really exciting news, I was expecting to read about mink tracks and feel gloomy that the time had finally come to kill a mink. But standing at my computer in the kitchen, I feel a thrill reading this. It is proof that as long as we protect them and give them a good habitat, the water voles will do the rest.

It's a really encouraging sign that we could really make a difference.

I rush outside across the bridge into the sunshine and past the herbaceous border, a sharp left, duck down, under and in between the trees and nettles and I am in the darkness, on the river bank looking back at our house. I stand on the willow trunk which juts onto the river like a jetty and grabbing the tether pull in the mink raft. I check the damp clay but it is just the same as the day I put it out weeks ago. Not a soul has been near it, except me rabidly checking every day.

Our mink raft has been just upriver from the house for weeks now and still no tracks, there have been tracks nearby but nothing on the raft at all, and we know that otters have been up and down the river. I try not to feel disheartened because Robin did say it would take about a month for all the residents to acclimatise to this strange thing floating in the river.

He was right, but water vole tracks weren't quite the result we were expecting.

I wonder how old that adult vole he refers to actually is and whether they still have time to breed this season. If they are spreading upstream, another litter here in the new location may mean that they spread still further in time for next year. The populations tend to contract in terms of space in the winter and expand again in the summer when they are breeding, so this could just be a natural part of that expansion and contraction but we didn't find any evidence of them there before.

It is an impossible dream but we don't have a shot of baby water voles – yet. I must start nagging Charlie …

Philippa: **Through the summer I had various discussions with Gareth the bee-keeper on the merits of this invasive species.**

What's the problem?

* An alien species that was introduced in 1839 and swiftly colonised riverbanks
* Each plant can produce up to 800 seeds which are exploded upriver as far as 7 metres and can be carried downstream as far as the river takes them
* It grows swiftly to great heights taller than Charlie's 6'4" and shades out our native bankside flowers
* Provides lots of nectar for pollinating insects thus luring them away from native wildflowers
* Grows on shoals and shallow banks and dies back in the winter, leaving those banks vulnerable to erosion in the winter floods

From his point of view Himalayan balsam is a freely available nectar rich plant for bees so can only mean good news.

The nectar rich flowers prove irresistible.

Himalayan balsam – an alien invader to seek out and destroy?

What to do about it?

* Pull it up before it grows too high or sets seed. This is easy (the roots are shallow) and therapeutic!
* If you plan to chop it down, make sure you do it before the end of June

At the end of another summer of pulling it up and feeling fed up as I see yet another bank of beautiful native wildflowers starved of light and space, I reluctantly admit that, yes, at least there is an upside to Himalayan balsam. I can't seem to beat it so at least the honey bees benefit and I will still be enjoying the honey on toast long after the flowers have faded in the winter.

Charlie: I've got into a great routine with the mother and cub. They're visiting the bridge holt every three or four days and their footprints in the sand let me know when they're in. The only problem with them is that when they do come out I often can't tell whether they're going upstream or downstream, which is frustrating as sometimes I chase and miss them.

Tonight I'm working from the kitchen so at least if I do miss them I can stay warm and comfortable. Camera assistant Ian is not quite so lucky. He's upriver in his car, watching remote cameras on the bridge holt to monitor the otters coming out and warn me if they're heading my way. You see, tonight I have a plan!

I'm trying to work out whether otters can smell underwater. According to science they shouldn't be able to. I have a hunch that they can. Otters hunt underwater using their eyesight and

Interestingly she never caught one of the young fit trout in her pond – they were just too quick for her. I would feed her at night, though, and to stop the rats pinching her food, I'd often chuck it into her pond. She'd then swim down and pick it straight up – how did she know it was there?

Tonight I'm hoping to find out. Upriver from the house, between it and the bridge holt in fact, is my new underwater rig. My brother Jeremy and I designed it for the very purpose of filming otters underwater. It is essentially just a

… they can lock on to the fish and chase them, even in pitch-dark, murky water.

whiskers. Their highly-sensitive whiskers can detect the movement of fish and they can lock on to the fish and chase them, even in pitch-dark, murky water. I have watched them do this many times, particularly when I had a tame otter. I would stand for ages at night watching her hunt fish in her pond. She would circuit the pond relentlessly and I would watch with a powerful torch. I could see the fish cruising around on the surface to avoid her. When she got too close to one it would bolt with a thrust of its tail, it was always at this moment that she would lock on to it and go for it. She would pursue the fish for a couple of laps before she'd have to grab some air, at which point she'd almost always lose the fish.

waterproof box with a camera in it. The camera looks out of a glass window in the side and there are various wires plumbed in to allow me to focus, zoom and record. In front of the box is a brick with a stinky dead trout tied to it. The trout is lit by a few infra-red lights but the image is dark and murky because infra-red light doesn't travel very far underwater. The underwater rig is wired to the kitchen where I'm sitting watching it on a monitor. I have several other shots wired to the kitchen so I can keep an eye on what is happening above water too. So now it's dark, about 9 p.m. and I'm waiting, drinking a beer and watching telly with iPlayer on Philippa's computer – the kitchen is my favourite hide!

9.15 p.m.

The phone rings. It's Ian. 'They're coming out now.' He stays on the line and gives me a running commentary of what's going on. I start to wish I was up there watching. 'They're rolling around together and playing in the sand,' he explains. Luckily Ian's recording it. Ian watches as the otters roll about grooming and playing for nearly five minutes. 'They're getting in the water.'

'Which way are they going?' I ask nervously.

'I'm not sure,' Ian replies. Without lighting the whole river it is very difficult to tell which direction the otters are headed. I wait and wait. 'They're coming your way!' I put the phone down and start to record. It's only a hundred yards from the bridge holt to the underwater rig so it won't take long. A minute later I see ripples appearing on one of my monitors. The shot I am watching is a wide one looking up the river. I watch as a large ripple appears at the top of my frame. More appear and the ripples move downriver, hugging the bank. The otters don't appear themselves; instead I watch their ripples work their way down past the camera on the opposite side of the river to the dead trout. 'They completely ignored it,' I curse to Richard, who's filming me.

The otters are now in the gap between the dead trout and the house, another 100 metre stretch. I have no cameras on it but I do have cameras on the weir, where the otters are heading. I frantically fumble around with the mass of cables that come in from the garden and spill out inside one of the kitchen units. All the cables are marked with what camera they go to. I find one which says 'under

weir', I plug it into my recorder and a picture appears of the weir and the step under the sluice gate – this is the otters' favourite route. I plug two others in, both different angles on the same route. I hit Record on all the monitors and look out of the kitchen window. Ripples appear in the reed bed opposite the house. I watch as they pass under our bridge and head off along the edge of the weir and downriver, totally ignoring their normal route and not appearing in any cameras.

'Typical!' I shout; otters can become incredibly frustrating. They have a real habit of doing the exact opposite of what you want and expect them to do. These two have just travelled the entire length of the river in my garden, including a stretch right outside my kitchen window which is floodlit, without me seeing them once. This is unsurprising but very frustrating, not just because you don't get the shot you need and want, but because the set-up involved takes days to rig and when the otters avoid every part of it, it smarts a little.

The otters have passed and won't be back tonight. Richard and I discuss tactics

An otter is an efficient hunter even in the darkness.

Halcyon River Diaries   201

and I explain that they'll be back at around 10 p.m. tomorrow. I say this because in my experience they will be back at around that time. Otters are as predictable as they are unpredictable. To my knowledge their timings on our river are dictated by where they slept during the day and where they've decided to sleep the next day. I rarely see otters travelling downriver more than an hour after dark because they do this after spending the day in the bridge holt which they generally leave an hour after dark and then it's only a few minutes to my house from there. When they're heading upriver it's different. I don't know where their holt is and they seem to appear at all times – 9 p.m., 3 a.m., sometimes even 10.30 a.m. However, in my experience, if they go down early one night, they come back up fairly early the next night. So I'm placing my bets on 10 p.m. tomorrow.

9:45 p.m. The following night

The otters have just come over the weir, once again avoiding all my cameras, including the new one I put out to catch them as they evaded the others! They're heading upriver to the dead fish.

I frantically de-rig the weir cameras from the kitchen recorders and plug in the wires that link to the underwater camera and others upriver by the trout.

I put all recorders into Record mode and watch the screens excitedly. A moment later ripples appear and I spot a bubble trail on the monitor. The otter is heading for the trout. The ripples get closer and closer. I turn my attention to the underwater camera, the shoal of minnows hanging around over the dead fish clears and in a murky swish the otter appears, grabs the trout and vanishes into the murk, taking the brick that the trout is tied to with it.

I scream and shout and leap about quite unable to comprehend what I've just filmed. Richard of course films all this and my shouts and screams bring Philippa down. We rewind the tape and watch it over and over. The shot is amazing but very brief, about three seconds. The otter appears from the right-hand side of frame, seems to nose its way up the trout until it reaches the head, then grabs it and swims off.

We watch the otters on the monitor for nearly an hour, hunting the pool where the dead trout was. They are very hard to see as they are either underwater or in the vegetation on the bank and I have no idea where the otter went to eat the trout. Eventually the ripples vanish and the river settles as the otters move on up towards the bridge holt.

The following day I put the footage onto the computer and watch it over and over again. I review it in slow-motion – this leads to a startling discovery. The otter comes into frame and immediately hits the tail of the trout with its nose; as it does this it lets out an air bubble from its nostril which it immediately sniffs back in again – is this how the otter smells underwater? What alerted me to the idea was a colleague of mine who recently filmed star-nosed moles smelling in a similar way – by putting out air bubbles and sniffing them back in.

I phone all my otter mates, some scientists, others obsessives. No one had encountered such a phenomenon before. Had we really just proved that otters can smell underwater? I think so, but obviously I'll have to spend several more years hanging out on rivers filming otters to find out for sure!

↓

The otter went directly to the trout and whisked it away.

Had we really just proved
that otters can smell underwater?

Philippa: September could mean a lot of hard work. The flood meadow is past its best, now it is just green. All the flowers are gone, much to the chagrin of the boys – even the Bee Orchid has withered.

But now is the critical time for management – the hay cut is vital if we are to restore the diversity of species. I have managed to persuade Colin from up the road to bring down his tractor and take the top off for me. It's not going to be straightforward – the ground is full of bumps and ants' nests and the thatch is just so dense that it will be hard to chop it at all. But Colin has promised to come and try. He has helped me with the garden since we moved in and has been a gardener since he was a boy. I wonder if he thinks I'm bananas, but if he did he would never say. He quietly assists the plants and the place through the seasons with no fuss and just the odd joke whilst we flurry around moaning, panicking, running and stressing. I have never seen Colin stressed: he never complains, he just is. Fit as a fiddle even though his hair is white, he can dig all day. There aren't many people left like Colin in the world, he is very special.

Sure enough on Saturday morning his little old red tractor, which looks more like a museum piece, comes through the gate, a delicate plume of smoke from her chimney announcing her arrival. She is old but so sweet, the engine barely making a noise. She is every boy's dream and Arthur is in love. Between us we rig the mower on the back, bolt it onto the second-lowest setting and Colin sets sail around the paddock.

Although he isn't keen on the noise and nervous of being run over, the tractor is like a magnet to Arthur, he can't help but watch. I bite my thumbnail, nervous about whether this will work – if it doesn't, I will have to do the paddock and the other bit by hand. That will take days which I don't have; it doesn't bear thinking about. But I am committed now and can't leave rank grassland where there could be precious flood meadow. I realise that I still have lots of old tractor grease on my thumb from adjusting the mower. I spit it out which Arthur thinks is hilarious.

As Colin sails around, the grass behind him falls. It is so thick but as we wade in to the mown sward, we realise that actually his old tractor has done her job beautifully. We put the mower onto a lower setting and

we would have no butterflies, no moths, no bees nor any of the other insects and creatures that depend on them. For me the thrill now is not just knowing what a big difference we are making by saving every little bit of flood meadow but also knowing the implications this has for the rest of our riverside wildlife.

Hay making is another connection with our past, another connection with a way of doing things that worked with the natural world – watching the swathes fall is deeply satisfying.

The tractor goes around and around, and now it has a new driver. Arthur has overcome all concerns about the noise and his smile is wider than the mower. He is in heaven. Hay making is something we were all designed to appreciate …

… Now there is just the raking to do …

the result is even better. There is still some thatch lying down but an afternoon with the strimmer should get rid of the worst of it. I feel a great sense of relief, lighter somehow as if it was me having the haircut rather than the meadow. Next year I will count the species again – in the new open conditions the more delicate ones will have had the space to grow. In each subsequent year there will be new ones and the flood meadow will have improved.

To so many people this is uninteresting, these are just plants, but without them,

Dragonfly wingbeats

Charlie: Dragonflies are incredibly agile fliers – they are able to perform aerial manoeuvres that are so fast they are often invisible to the human eye.

Their wings are a feat of design and engineering – they are very light and thin but incredibly strong and robust. The wings are powered by a large set of muscles in the thorax of the dragonfly which beat them so fast that to us they are just a blur. When dragonfly expert David Smallshire visited to tell us about dragonflies I asked him how many times per second the wings beat. He wasn't exactly sure but suggested around thirty times. I decided to investigate.

I managed to get my hands on a Photron. This specialist hi-speed camera can film at astonishing rates – upwards of 10,000 frames per second (or fps)! In order to shoot in slow motion you need to increase the number of frames the camera takes every second. Normally a video camera shoots at twenty-five fps. This is enough to create a smooth moving image when all the frames are played in succession. If you increase the number you shoot to fifty fps you double the amount of frames taken within the second. So the action that occurs in one second takes two seconds to play back – it's being slowed down. Shoot at 2,000 fps and you're slowing the action down eighty times!

My feeling was that if I really wanted to slow down the movement of the dragonflies' wings, I would have to shoot at around 2,000 fps. This would allow me to see each wing beat and count them accurately. But the problem is that the faster your frame rate the more light you need. As a result our first attempts failed because we simply didn't have the light – we had grey instead of sun and after a few days of trying we had to send the camera back to its owner with nothing to show for it.

We got it back a few weeks later and this time the sun was out. After spending the afternoon in the pond we began to get some decent images. The most numerous species of dragonfly in the pond is the common darter, a smallish, red dragonfly. This was to be our guinea pig. Our aim was to get good clear shots of the dragonflies in flight – much more difficult than I'd expected. They fly very fast and I was struggling to keep them

↓

The common darter is not the easiest species to film.

The dragonflies had been slowed down so much they almost didn't look real.

↓
Scientists agreed the dragonflies beat their wings around thirty times per second.

in frame and in focus. To make matters worse I was using a Nikon lens. I don't normally use Nikon lenses and they focus in the opposite direction to other lenses. This means that my brain was working overtime trying to focus backwards. We filmed the darters over several days and when we thought we had enough, took the footage to the edit suite to inspect it.

The footage looked amazing on the large screens in the edit; the dragonflies had been slowed down so much they almost didn't look real. We isolated an eight-second clip of dragonfly footage, counted the number of times the wings beat and did some calculations. Information from scientists suggested a

wing beat rate of 35–40 beats per second for the common darter – looking at our footage we were counting 80–100! We got very excited by our new discovery – not only had we got some great footage but we'd completely rewritten the science. It was a few days and a little more calculating before we realised we'd been completely wrong. Further investigating revealed the actual wing beats per second to be 55 – many fewer than we had previously thought, but still 15 per second more than scientists had suggested. We examined seven different shots and discovered that every single shot had the same rate of 55 per second – the new speed of dragonfly wing beats!

Ten great places to see ...

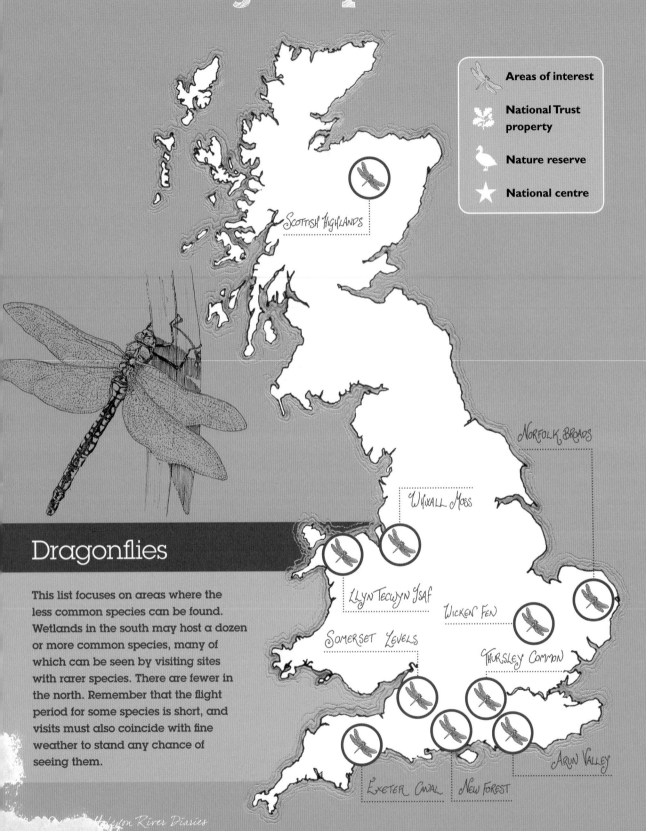

Areas of interest

National Trust property

Nature reserve

National centre

Scottish Highlands

Norfolk Broads

Whixall Moss

Llyn Tecwyn Isaf

Wicken Fen

Somerset Levels

Thursley Common

Exeter Canal

New Forest

Arun Valley

Dragonflies

This list focuses on areas where the less common species can be found. Wetlands in the south may host a dozen or more common species, many of which can be seen by visiting sites with rarer species. There are fewer in the north. Remember that the flight period for some species is short, and visits must also coincide with fine weather to stand any chance of seeing them.

Halcyon River Diaries

Scottish Highlands

The three Scottish specialities, northern damselfly, azure hawker and northern emerald – together with common hawker, downy emerald, golden-ringed dragonfly and white-faced darter – can be seen at lochs and bogs. Most of these can be found at Coire Loch in Glen Affric.

Whixall Moss National Nature Reserve, Shropshire

This peat bog is the most accessible location for the white-faced darter in England. Emerald damselfly and black darter also occur in the restored wet areas.

Llyn Tecwyn Isaf, Wales

This lake in western Snowdonia has small red damselfly, hairy dragonfly, common hawker, keeled skimmer and downy emerald, the last at its only known location in North Wales.

Norfolk Broads

The endangered Norfolk hawker flies in spring, typically along ditches rich with water soldier. Upton Broad and Marshes Norfolk Wildlife Trust Reserve is a good place to find it, together with variable damselfly and hairy dragonfly.

Wicken Fen, Cambridgeshire

This National Trust reserve also hosts the National Dragonfly Centre, run in conjunction with the Dragonfly Project and the British Dragonfly Society. It's a great place to learn about dragonflies, as well as to look for variable and small red-eyed damselflies, hairy dragonfly and scarce chaser.

Thursley Common National Nature Reserve, Surrey

The Moat Pond has small red damselfly and downy and brilliant emeralds, while the bog and stream to the east have beautiful demoiselle, golden-ringed dragonfly and heathland species such as common hawker, keeled skimmer and black darter.

Somerset Levels

Hairy dragonfly is widespread in spring in the ditch systems, but variable damselfly is rarer and scarce chaser even more so. white-legged damselfly can be found on the rivers, red-eyed damselfly are on the larger drains where water lilies abound and huge numbers of more common species breed in the flooded peat workings.

Arun Valley, West Sussex

Look along the riverbank above New Bridge, Billingshurst and Stopham Bridge, Pulborough for hairy dragonfly, common club-tail, brilliant emerald and scarce chaser. Further down the valley, variable damselfly is found at Pulborough Brooks RSPB reserve and Amberley Wildbrooks.

New Forest, Hampshire

Streams here host our most important populations of the endangered southern damselfly. Crockford Stream has lots of them, plus small red and scarce blue-tailed damselflies. Look for the latter in open areas where livestock cross. Similar species can be found in wet areas of the Isle of Purbeck in Dorset, the East Devon Pebblebed Heaths and Dartmoor fringes in Devon.

Exeter Canal & Exminster, Marshes, Devon

Look for hairy dragonfly and white-legged, red-eyed and small red-eyed damselflies along the canal either side of the Countess Wear swing bridge. Many of the grazing marsh ditches east of Exminster also have hairy dragonfly, while a few may have scarce chaser.

Philippa: It is late September, a good few months after the start of our mink monitoring programme. Upriver near Chris's raft, there have been more tracks – water vole tracks.

It is a bright, warm, blue day and the voles have spread well over five hundred yards up the next ditch. But the best is yet to come – as well as adult tracks there are others, possibly babies. There is only one way to be sure.

The children are back at school and so we only have Arthur, which makes life a little easier. Charlie placed the apples in the ditch and set his cameras early in the day – this vole project has used most of the apples on my tree. By late afternoon Arthur and I go to find out what he has filmed.

Sitting beside the ditch stuffing ourselves with the very last blackberries, we watch the small video screen and see the by now familiar shape of an adult water vole but then behind comes the identical but smaller shape of a baby.

Arthur is more interested in finishing off the blackberries. He's seen water vole

pictures before and at three is slightly too young to see what all the fuss is about, but even he falls for the baby. For me it couldn't get any better. The water voles are not only expanding but are also breeding and we have the proof.

Just a week later, more tracks, only this time it is on our raft and the tracks are mink. We know what we have to do and so with heavy heart we replace the clay pad with a trap, something which we have not had to do so far.

I wonder if we will finally have to shoot a mink. I feel sick at the thought of it. We wait, we check the raft morning and night and there is nothing, no sign, no animal in the trap, no mink tracks on the raft further upriver.

There are other tracks, bigger ones, and we know who they belong to.

One night we are sitting chatting in the kitchen and outside the water starts to move, large ripples. None of the cameras are charged.

'Bloody typical' Charlie grumbles. but it means this moment will be ours. We usher the dogs inside and sneak out, careful not to make a noise. We are familiar enough with the ripples to know that we have a few moments before the otter appears outside the house. We stand, hand in hand, watching the water. Then comes a sound almost as familiar to us as our own baby crying, but one that we haven't heard in ages. It is the high-pitched whistle that young otters make when they are travelling with mum.

They appear just in front of the house, skirt around the mill pond, under the bridge and then back again. The cub

↓
Clear mink prints in the clay pad.

The water voles are not only expanding but are also breeding and we have the proof.

jumps off the sluice gate and returns for another play in the pond before being moved along by its mother. Seeing a wild otter is so rare that sharing more than a moment with one is a blessing, sharing it with Charlie, who feels exactly the same way, only more so.

Over a week passes since we saw our tracks and there is no more sign of mink. There are a few rafts between ours and the vulnerable vole population at the top of the river and there is no evidence that the mink has gone any further up towards them. It is a critical time, if we miss the signs the vole population will be in trouble. But there is nothing and I finally begin to relax into the hope that the trigger will not be pulled.

I know that come the winter water vole territory will naturally contract a little again, this is well-documented, but to have them spread this far and still be breeding this late into the summer means that even accounting for that natural contraction, the water vole population is spreading overall. Perhaps we have made an unexpected discovery, could it be that in a world where every forecast is a gloomy one the forecast for the water vole is sunny?

Almost ten years after we saw our first otter exploring his new territory, we are perhaps now witnessing the consequence of the otter's return. Could it be that, since the otters are coming back to our rivers, they are outcompeting the mink and the natural balance is being restored, and the water voles are in with a chance?

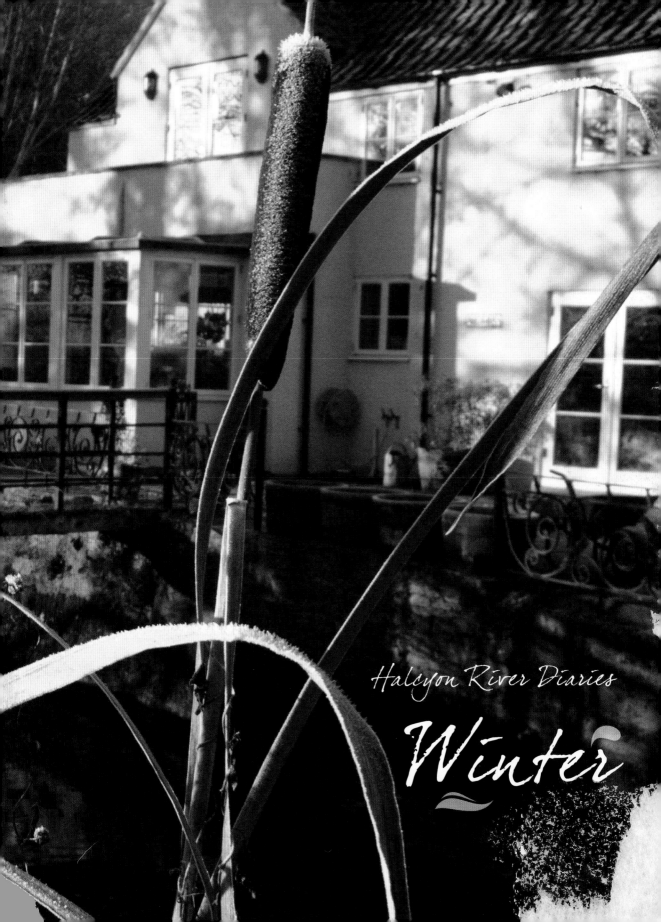

Halcyon River Diaries

Winter

Philippa: At last, eggs through the winter. As I watch them now the ducks are happy foraging in the stream, so engrossed that sometimes all I can see are fluffy bums. Their heads are under the water and they are busy dabbling, stirring up the mud with their feet and seeing what comes up. Their tales waggle from side to side and they are a picture of contentment.

Our wild duck orphan has had his first moult and it turns out that he is not, as we had presumed, a female but instead a rather handsome, if small, male. He has no intention of returning to the wild any time soon, although he has the freedom to leave whenever he chooses.

Now he is foraging alongside one of the large, white Cherry Valley ducks. There are one drake and six females in total and he is tiny by comparison – they are at least twice his size. Now that he has his adult plumage I wonder when he will realise the difference and when his masculinity will really kick in. That will be amusing to watch …

The original Khaki Campbell, who I think of as simply 'small', has regained all her confidence, perhaps she has forgotten the trauma of the fox disaster. Although she will always be 'small' to me, now they are all fully grown it is often hard to tell them apart.

The Khaki Campbells are the best layers – they can lay over three hundred eggs a year. When, in the depths of winter, the day barely glimmers into action and there is just a brief period of light in the middle of the day, the chickens give up egg-laying completely. I don't blame them – if it wasn't for Christmas, my productivity would cease altogether as well. Apparently the ducks won't stop laying, and so now we can look forward to fresh, free-range eggs all year round.

Their eggs are white, about the same size or a little larger than a chicken egg, with deep orange yolks. Much is said about the difference between duck and chicken eggs and which ones are more suited to which dishes, but we interchange them regularly. They work well in most dishes – they are lovely as eggs Benedict for a special breakfast but are also especially good for baking (apparently because they contain more albumen).

If I have a glut of eggs, I whip up a few of them – and make a cross between a meringue and a macaroon. They work really well as a dessert base for fruit and cream, and are really nice with blackberry sorbet on top (see page 196).

Indian runner ducks are great egg layers and slug eaters.

Dutch duck egg cakes

When I first began searching around for duck egg recipes I found this one on the internet and played around a bit with it – it is really easy too (so thanks to DJ in Scotland). This makes enough for five biscuits.

1 large duck egg | 60g sugar | 60g flour | ½ teaspoon baking powder icing sugar for decoration

Whisk up the sugar and eggs until they are foamy. Slowly add the flour and the baking powder, stirring all the time until you have a light, smooth, runny batter.

Line a baking tray and then dollop the mixture on in 1 tbsp portions, leaving plenty of space for them to spread.

Bake at around 200°C/400°F/Gas 6 for 10 to 15 minutes until golden brown.

Just like the ducks, a finished stack of these, lightly dusted with baking sugar, always brings a smile to my face.

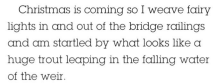

Philippa: I know that winter is here because my feet are cold. I'm no good at winter. The river reflects grey sky. The beautiful beech tree outside by the weir has lost the last of its orange display and everything in the garden is dead or dying. The aquarium has been emptied, all its residents returned to their wild home. Many of the cameras that were lining the river have been brought back. I clear the dead plants from the vegetable garden and mud is everywhere. It feels like the whole world has sunk into gloom.

But if I stop and make the effort to look there is still life all around the river. Swallows and swifts disappear but the kingfishers stay, taking time to perch on the bridge, as do the dippers. Even in the cold, our river neighbours stay. Ducks loiter outside the house, Mr and Mrs, our usual couple, still together after another year rearing ducklings.

Christmas is coming so I weave fairy lights in and out of the bridge railings and am startled by what looks like a huge trout leaping in the falling water of the weir.

Like many we feed the birds, and those that were spread through the garden and fields come together spending their days right outside the kitchen window: great tits, goldfinch and long-tailed tits, noisy and arguing, bringing life to the patio.

The long nights often mean that otters are more active in the winter. By four o'clock we can look out and see that they are already up and hunting. Winter is the time when older cubs still hang out with their mothers. Whistling their way up and down the river, not babies any more – finding their feet, hunting their own fish – but still relying on mum for security.

Frosty dawns bring to life the spiders' webs all along the bridge, delicate white spangles more beautiful than any Christmas decorations I could add.

On cold mornings mist rises magically from the weir pool, dramatic against the naked limbs of the trees, the unsure sun rising behind them.

We wrap up warm and walk along the river, across the frost-hard fields. We find ourselves at the spot where we'd picnicked in the summer. What a different place. Its banks stark and sparkling, its water hard and cold.

For the tenth year we drag a Christmas tree across the bridge, terrifying a moorhen

Frosty dawn brings to life the spiders'
webs all along the bridge, delicate
white spangles more beautiful than any
Christmas decorations I could add.

who scoots across the water back to the safety of her dilapidated reed bed.

If we are lucky we may have snow again this winter, and as they open white-scened Christmas cards, the children talk about the tracks they saw last time and those they will follow this time, they talk of mink and otter and vole and I realise how much they know about this place.

Thanks to the river the inside of our home springs to life. The wood-burning stove burns brightly driving the chill out, with wood that the river brings us through the year; every time the water is up the logs come floating down and we simply collect them at the bottom of the weir, and chop and dry them. Our own personal delivery service.

I sit and stare at the flames. I know that it is really just a few weeks until the first spring plants begin to stir in the soil. My feet are warm again.

Useful websites

www.anniehallspoultry.co.uk
Annie Hall is a great source of chickens, ducks and drakes, and everything you will ever need to look after them – along with a smile that never fades.

www.arkive.org
This is a family favourite, it appeals to all ages and has an incredible collection of facts, images and videos which is constantly being updated. An unmissable window on the world of wildlife. There are also great educational games for children.

www.basc.org.uk
The British Association for Shooting and Conservation – it may seem at odds to some that the same body that promotes shooting also promotes and practices practical habitat conservation but this organisation has been invaluable in helping with our water vole project and similar projects all over the country.

www.btcv.org
This organisation provides support for conservation projects all over the UK. If we have inspired you to take action – to improve the habitat for your local water voles, for example – then these are the people to contact. You can do anything from the odd day to a whole holiday.

www.butterfly-conservation.org
A wonderful website that I log on to frequently with a cup of coffee, just to look at the pictures and learn a little more. Butterflies and moths surround us all and yet we know so little. This website will change all that.

www.dragonflysoc.org.uk
The British Dragonfly Society aims to promote and encourage the study and conservation of dragonflies and their natural habitats, especially in the UK.

www.naturalengland.org.uk
The body which advises the government on the natural environment and is responsible for sustainable stewardship of land and sea. For us, that means the organisation we apply to for licenses to work with otters, kingfishers and water voles, but their work is far reaching and anyone interested in conservation should be familiar with it. The website is the best place to start.

www.environment-agency.gov.uk
The government agency which is responsible for our rivers and flooding. These are the people who worked with us on electrofishing and eels, they are running a programme to save the native crayfish and are responsible for protecting our rivers from pollution.

www.firebellystoves.com
These stoves, designed and built in the UK are so efficient that we haven't had the central heating on at all this winter.

www.heritageireland.ie
The body entrusted with the protection of Ireland's natural and built heritage.

www.craigjoneswildlifephotography.co.uk
Craig Jones runs various workshops for people of all ages and abilities who are interested in photographing wildlife, with workshops on dippers and water voles.

www.mammal.org.uk

An organisation devoted to the study and protection of all British mammals – a great starting point for the latest information and a forum for discussion, it is worth checking their website regularly.

www.otter.org

The home of the International Otter Survival Fund in Skye. It was founded by Grace and Paul Yoxen but it protects otters all over the world. The website is a great place to find out more about their work, and about otters – there is much to learn here. The children really like the activities and photos on the kids' page – and if you want to adopt an otter as a present you can do that too.

www.plantlife.org.uk

Plantlife is a wildflower, plant and fungi conservation charity. They gave us invaluable help when it came to our flood meadow. Their contribution to the world of conservation cannot be underestimated and their website is an inspiration.

www.rivernet.org

The European Rivers Network seeks to promote the sustainable, wise management of living rivers.

www.rspb.org.uk

A heavyweight charity who lead the way in education and lobbying. By securing a healthy environment for birds they ensure that many other species are safe too.

www.shetlandotters.com

Terry and John, your friendly guides, will show you otters in their natural habitat. Otter watching in Shetland is civilised as it is done during the day and the chance of seeing one here is higher than anywhere else in Britain. This leaves the evenings free to boast in the bar.

www.snh.org.uk

Scottish Natural Heritage is the body responsible for looking after the nature and landscapes of Scotland. In a similar way to English Nature they provide grants and licenses, carry out research and protect and care-take sites; working in partnership with many other organisations.

www.sttiggywinkles.org.uk

A huge and wonderful wildlife hospital, for any wild animal which is in trouble, these are fantastic people to call for help and advice.

www.watercress.co.uk

A great place for all things watercress or to find out what's on at the next watercress festival.

www.wildlifetrusts.org

These are the people to get in touch with if you want to help your local wildlife, it is a voluntary organisation with local branches and it achieves an enormous amount. Their website is full of information on UK wildlife and how you can help.

www.wildlifewatchingsupplies.co.uk

They have everything you could ever dream of when it comes to camouflage, from lens covers and clothing to hides.

www.wwt.org.uk

The Wildfowl and Wetlands Trust is a conservation charity close to our heart, and the modern dream of the visionary Sir Peter Scott. We were honoured to open the wonderful kingfisher hide at Slimbridge. There are nine UK wetland centres – all offer wonderful days out, whatever the weather, and most importantly the chance to get close to wetland environments and understand them.

Index

ACKNOWLEDGEMENTS

Philippa, Charlie, Fred, Gus and Arthur would like to thank the team who have made this year so amazing. They are: Trevor Dolby, who would love us to write about his talents endlessly. Nicola Taplin, managing editor, who said keep it short, sorry Trevor. Craig Stevens for his design talents and Nick Heal for project managing the book, with memories of a hilarious evening in which we attempted to work with him, film otters and look after three kids – he wasn't at all fazed. The team at Butler, Tanner and Dennis. Vickie Boff, Jessica Axe and the Preface sales team. Charles Armitage for wisdom and inspiration. Hilary Knight for love, support and supreme efficiency. Jake Biggin for his beautiful drawings. Our production team for the television programme. Richard Taylor-Jones for his considerable talents at stills photography, keeping Charlie's head an appropriate size and Philippa sane, and his friendship. Ian Llewellyn for his hard work and wonderful images. Stephen Hamilton your sound and logging were wondrous. Sam Organ for courageous calm in the face of unbelievable adversity, you inspired us. Andy Chastney and Nigel Buck for your talents and dedication. Carolyn Naylor and Sarah Tooley for your dogged determination and endless flexibility. Annie McGeoch for her number crunching.

At the BBC Jay Hunt, Cassian Harrison, Jo Ball and Mark Bell. At home, the friends and family that have helped us with children and practicalities to enable us to work, particularly Dave and Laura. All the contributors who helped with information and filming days: Dominic at Plantlife, Richard Fox at Butterfly Conservation, Watercress Wendy at Mustard PR, Avon Wildlife Trust, Robin Marshall Ball and his wife Pam, our mink raft team, Craig Jones, Phoebe Carter, Jeremy James, Annie Hall, Dave Smallshire, Gareth Baker, Richard and Stephanie Horton, John and Sue Bailey, Diane Brain, Dave Elliot, Jamie McPherson, Caroline Michie and the team at the Environment Agency, especially Peter Sibley.

PHOTOS

© Nick Heal: page 192; © iStockphoto.com/Martin McCarthy: page 53; © Craig Jones: pages 108 (bottom), 198, 213; © Ian Llewellyn: pages 9, 10, 11, 21 (top), 38, 55, 62, 119, 125, 135, 140, 141, 171, 185; © Warwick Sloss: page 12; © Richard Taylor-Jones: pages 7, 15, 19, 66, 71 (bottom), 103–104, 106, 120, 124, 152 (top), 153, 159, 160 (left), 162, 181, 187 (top), 188, 193. **All other photos are copyright of the authors.**